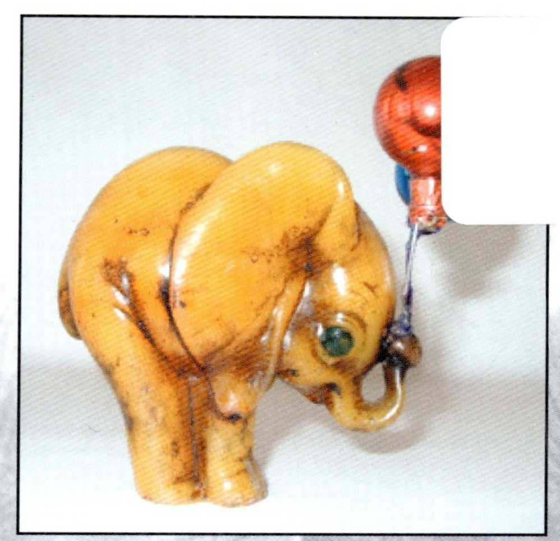
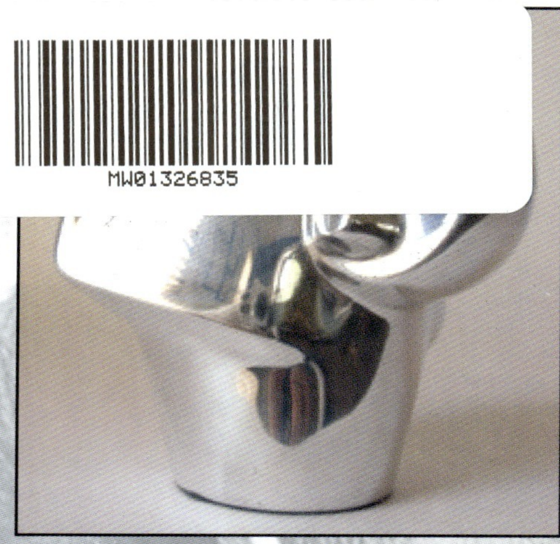

Everything Elephants

A Collector's Pictorial Encyclopedia

Michael Don Knapik

4880 Lower Valley Road, Atglen, PA 19310 USA

Dedication

I dedicate this work to all of my fellow elephant collectors, and, especially, to my wife, Terese, who not only helped create this book, but lives with my elephant-collecting obsession. I Love You!

Library of Congress Control Number: 2001098926

Designed by Bonnie M. Hensley
Cover design by Bruce M. Waters
Type set in BankGothic Md BT/Korinna BT

ISBN: 0-7643-1494-7
Printed in China
1 2 3 4

Published by Schiffer Publishing Ltd.
4880 Lower Valley Road
Atglen, PA 19310
Phone: (610) 593-1777; Fax: (610) 593-2002
E-mail: Schifferbk@aol.com
Please visit our web site catalog at **www.schifferbooks.com**
We are always looking for people to write books on new and related subjects. If you have an idea for a book please contact us at the above address.

This book may be purchased from the publisher.
Include $3.95 for shipping.
Please try your bookstore first.
You may write for a free catalog.

In Europe, Schiffer books are distributed by
Bushwood Books
6 Marksbury Ave.
Kew Gardens
Surrey TW9 4JF England
Phone: 44 (0) 20 8392-8585; Fax: 44 (0) 20 8392-9876
E-mail: Bushwd@aol.com
Free postage in the U.K., Europe; air mail at cost.

Contents

Acknowledgments

I wish to thank so many people for helping me with this endeavor; it was truly an *elephantine* task (arr, arr). Over and above the people whose elephant images I used throughout the book, and who are either mentioned in the attribution or choose to be anonymous, there are several particular people I want to thank as follows.

First, my spouse, Terese, who had to endure not only a collecting obsession, but a book-writing obsession as well. (You fellow writers out there know what I am talking about!) Terese helped me on several photo shoots of major collections, provided assistance creating captions for the images, and proofread the text. So this is just as much her book as mine.

Next I want to thank certain people who provided several images and a lot of information: Adele Verkamp, Ted Capell, Janet Wojciechowski, Michael Par, Robert Selman, May Marwick, and Jane Chapman.

Finally, I want to thank all the particular artists, dealers, auction houses, and individual collectors who provided one or more images for this book. Even though I asked many people for permission to use their images, at the end, many images had to be eliminated from consideration for this volume. (In addition, many who contributed images wished to remain anonymous.) However do not be too disappointed if one or more of your images and information did not get into this volume; follow-on volumes of *Everything Elephants* will include your contributions.

Foreword

I collect elephants. Or. I'm an elephant collector. To those of us who do collect; and for those of us who are collectors, we know the meaning of those two phrases. For whatever reason, you make your hobby: "*elephant collecting.*" I share your interest. I have been involved in this diversion for some 30 years. I set out collecting "collector books" on whatever topic just for a reference library. Now the books in my library, almost equal the size of my elephant collection in number. This book is overdue. Mike has put together an elephant collector's guide to aid the starting collector or the "seasoned" collector, like no other book on the guide market. Future volumes will include more elephants and in-depth treatments of many particular categories like logo histories, trademarks, and the uses of the elephant image in many applications as well as — well — more elephants. These elephant reference guidebooks of which this is the first, will start a revolution in elephant collecting. Giving buyers and sellers alike an aid into values and information, the timing is perfect!

Randy V. Jackson
Newelephantman

SECTION 1

ABOUT THIS BOOK

INTRODUCTION

I intend this book to be the first in a series of works comprising a comprehensive compendium of information about elephant collectibles and collecting. "Elephant collectible" means something that takes the shape or form of most or all of an elephant, or has an elephant (image or sculpture) upon it as an embellishment. I hope this book presents as wide a range of elephant things as you are likely to ever find, anywhere!! So, first, let me tell you how to contact me if you have questions or suggestions for follow-on volumes and other activities.

My e-mail is: conscioussystems@mindspring.com.

I wrote this book because I wanted a broad reference on elephant collectibles, and could not find any in the market. The only other (2) books I found, as good as they are, do not come even remotely close to showing the breadth and depth of elephants collectibles I wanted. One is *An Enchantment of Elephants* by Emily Gwathmey, published by Clarkson N. Potter, Inc. in 1993 and now out of print. It is a wonderful book that describes the lore and myths of elephants and presents many images of elephanteria ephemera, but it is rather small at 80 total pages. The front cover is shown in the books category of Section 3. The other book is *Elephant Ancient and Modern* by F. C. Sillar and R. M. Meyler, published by Viking Press in 1968. It delves more deeply into the use of and description of elephants over the millennia, and provides examples of ephemera and literature related to elephants.

But I wanted more, and in the format of most other collectibles books; that is, hundreds of pictures of collectibles with as much description as possible, including valuations. I did not want to buy a couple hundred books on all the various types of collectibles in the hopes of finding a few elephants in each one!

So I decided to write this book- and follow-ons as necessary, to further my dream of a major, comprehensive work, whether that takes 2 or 10 volumes, plus a planned web site for tens of thousands of images and information! Of course, no book of this sort is absolutely complete. That would take an indefinitely long time to create, as new elephant stuff is always being produced, and the book would be too large and too expensive to produce; therefore it would be too expensive for most people to buy!

For example, as I went about collecting subjects and images, most times I would arrange to photograph a collector's collection. As I have seen perhaps tens of thousands of elephants, I assumed that I must have seen them all by now!! Boy was I wrong … every time! I am amazed at how many elephant things there are. In my estimation, there are surely hundreds of thousands (perhaps on the order of millions) of different elephants, including color variations of similar "models." So, this book is a start a pictorial catalog, if you will, of as many different types of elephants and related information on each collectible as I could gather in a reasonable amount of time.

This book includes about 1000 images picked from about 5000 to which I had access (either my own or others), so you can see that in this one volume, there can be only a very small percentage of possible elephants. I am sure I also missed manufacturers and artists in this first volume. Nevertheless, it is breadth of coverage I am shooting for foremost; I want to show as many different *kinds* of elephant collectibles as possible. Hopefully show many that you have not seen before!

That said, this book's depth of coverage is mostly in the Figurines category, simply because that is what most elephant artisans have created when their subject matter was elephants, and probably what elephant collectors collect most. Consider a category like snuff bottles: while there may be 325 different elephant snuff bottles, I could only put but a fraction the ones I have seen in this book.

Concerning the categorizations I used, some categories could be merged into other, more inclusive categories (and vice versa), but here is my rule of thumb. If there were more than 1 or 2 images of a type of elephant collectible, or if I thought that, in the future, I would find several other examples (for follow-up volumes of this work), it likely got its own category. I am sure I will add more categories in the future and any suggestions would be welcome. You can also find some elephants that could go into other, or more than one, category. For example, a political ashtray can go into Ashtray category or into the Political/GOP category. Future works will contain more figurines of course, but I will concentrate on getting several more examples and associated information into each of the other categories! In this volume I wanted to show as many different elephants from as many different manufacturers as possible and that meant concentrating on the figurines proper.

In categories where I put only a few examples, you can check Section 4 for further references and resources. Use the images and information in this book (even the lightly-covered categories) as examples to compare your own elephants to, or when shopping, to verify a piece, or even to get an idea of what a particular manufacturer's or artist's works look like in general. Many captions have a direct reference to a web site with more information or other elephants.

I know, I know, you are certainly going to look for your collectible at once - whether you are a collector or an artist. Although you will probably find some, perhaps many of yours in this book, it is just as likely that you will not find a particular one

A great majority of my own elephants are not in this book!! But, if you are a collector, or a real elephant lover (or both), you can use this book as a guide to the types of elephant things that are out there celebrating one of the coolest animals to walk the planet.

I envision this book as a handy *large* list of the *types* of elephant collectibles that exist and therefore, that you can collect. It is also a reference to a majority of the well-known manufacturers of elephants, and inclusion of many of the most recognizable, "important," and desired elephants. If you are an artist or manufacturer that has created elephants not in this book, or have many more than I have shown, please contact me. I am collecting images and information for the next book.

Now, considering that I said: "...this book presents as wide a range of elephant things...", I want to qualify that a bit more. I excluded some things that perhaps some collectors would consider an elephant collectible. For example, some furniture has legs that are in the gross shape of an elephant trunk or even an elephant leg/foot. Or some vases have what are called elephant handles, which means, typically, that the handle is merely in the shape of an elephant's trunk. I think many of these are incidentally elephant things, and not truly elephant collectibles. But of course, that criteria creates a fine line, and you may find exceptions herein.

I also excluded some items that may have an elephant on it, but which elephant is only a very minor component of an image or figure. For example, a picture of Noah's Ark, with perhaps twenty pairs of animals, including an elephant pair, is not necessarily an "elephant collectible" in the sense I intend for this book. So I am trying to say that this book presents things that are *primarily* elephant-related. The elephant must be the focus or predominant, easily recognizable motif of the collectible.

In addition, I had many more images to consider than I knew would fit in this book. One of the criteria for eliminating images to include in this first book, was whether it was an inexpensive example in a category that had many other choices. If that happened, the inexpensive ones got tossed, unless it was fairly unique in material or design, or the only example of a manufacturer or artist's work.

Finally on this subject, there is a period of time, wherein many elephants were created but, other than coins, there is no representation in this book. That period is approximately the 17th century and before, especially the 11th through 15th centuries, from Southeast Asian and India. After scouring scores of fine art auction catalogs from Sotheby's and Christies, I realized that to do justice to the hundreds of examples of paintings, carvings, and metalwork elephants I saw, those categories would require volumes of their own; they may fill a goodly portion of one of my next volumes of this *Everything Elephants*. For example, there are many steles (slabs of stone wherein carvings are rendered) made of sandstone and other rock, depicting Ganesha throughout the ages mentioned. Most of these works are in the $5,000-5,000 range in value, and are therefore a bit more difficult to come across in the normal course of research. So I decided my initial coverage for this volume would focus on those elephant items that are a bit more common and collectible by a great majority of collectors. But I do not want to underestimate the importance of the earlier works and will do them justice later.

Why I Wrote This Book

Collecting elephants is a passion for me! Until this book, there was no one place I could turn to find out about as many elephant collectibles as possible. By the way, my first elephant, I believe, was actually something my mom bought, probably in the late 50s or early 60s. It looks similar to the cobalt blue Limoges elephant shown in image 03.053.067 in Section 3.

Most other collectible books concentrate on a single type of thing (like bookends or salt & pepper shakers), or the things produced by a single manufacturer or artist (Royal Doulton or Meissen). For a topical collection (like elephants) that cuts across manufacturers and types of things, as well as materials, it is difficult, if not financially prohibitive, to get all the books necessary to figure out what has been produced with respect to elephant collectibles. An elephant collector like myself would have to get books on:

- Royal Doulton
- Cybis
- ephemera
- porcelain/pottery/china
- bronze works
- rock and mineral carvings
- Meissen
- Ronson
- L. V. Aronson
- Bakelite
- celluloid
- Rookwood
- Royal Worcester
- bookends
- salt and pepper shakers
- lighters
- toys
- Steiff
- several stamp catalogs
- several coin catalogs ranging over centuries
- etc., etc., etc.

All that and much more just to pick out the one or two pages in each that have elephant images! This book cuts across all categories, combining them, while focusing on a single collecting theme — elephants!!!

Plus, I have added photos and information on elephants in my and others' collections, which are probably not in any book anywhere! So writing this book was a way for me to express my passion and bring together information that I want about elephant collectibles.

Organization

There are 5 major sections to this book:

1. This introduction, which tells you about the organization of this book, and how to use the book.
2. Elephant collecting from my personal perspective.
3. The "Elephant Encyclopedia" which contains a categorized, pictorial presentation of elephant collectibles
4. A list of resources and references that you can use to

find out more about elephants, including web sites, museums and other organizations, and of course, other books on collectibles.

5. An index.

Section 3 Organization, the "Pictorial Encyclopedia"

There are many ways to organize elephant collectibles, because the exemplars cut across so many facets of human experience and endeavor: art, war, utility, etc. If the artifact or material exists, you can be pretty sure someone has put an elephant on it, or made an elephant out of it to celebrate, express, or sell something.

The primary organization of the images in Section 3, is by alphabetical listing of categories, or types, of elephants. That is, I decided to categorize elephants into types of things like: bookends, incense burners, stamps, etc. Organizing by low-level type helps put some order onto the world of elephant collectibles. If I choose just a few high-level categories, say for example, Tobacciana, I would have put all kinds of things in it including: lighters, cigarette premiums (felts, cards), humidors, pipes, pipe holders, pipe tampers, matchbooks etc. But that would make for a category filled with lots of disparate collectibles. Instead, in my scheme, each of those lower-level things has a category of its own. This makes it easier for you to find what you are looking for and serves to distinguish and accentuate the variety of things elephant. Additionally, the Figurines category is further sub-categorized by major material (glass, metal etc.).

Note that some examples could go in more than one category, but I avoided that because it would be redundant and grow the book too big. So I picked one. However, the index helps locate specific manufacturers and artists.

Also realize that some categories have lots of elephants in them and others are rather sparse. That is a consequence of the commonality of some things versus others, and to what collectors have concentrated upon. For example, there are lots of elephant stamps and coins and bookends and figurines, but not nearly as many elephant clocks, fountains, or eggcups.

I tried to collect as much information as possible about each image. What I was able to amass was not only limited by time, but by the knowledge of the owners and the research resources I consulted. For each item, I tried to list: a brief description (including manufacturer or artist), size, year(s) produced or introduced, and approximate value. Even though I could not find all the information on some items, I did not think that was reason enough to eliminate the collectible from this book.

Value Considerations

Throughout this book, the values appear after most every elephant (some, like a image of a trademark per se, did not receive a value). Now a disclaimer on those valuations. As you know, values fluctuate over time, so that what appears in this book when published, can certainly be different than what appeared five years ago, or five years from now. The values given are for the piece as shown in the image. "Mint In Box" or specially signed versions of the same collectible often command greater prices. Second when a wide range is specified, for example $400-600, it *usually* means that I found a difference that is based on the venue of purchase. For example, at a flea market or auction (where the bidding is not furious for the item) you could pay half of retail or less, versus full retail at an upscale antiques parlor; that differential usually applies for all collectibles. Another reason is that there may be a variety of colors or other permutations for a given elephant, and some permutations are rarer than others. One example of this is the Frankoma elephant cup series.

Also realize that, except for elephant figurines proper, you are competing for a collectible with those collecting it because it is some other type of collectible, incidentally having an elephant motif. For example, a 1920 ashtray made by company XYZ, with an elephant on it is collected by ashtray collectors, elephant collectors, XYZ collectors, and perhaps art deco collectors as well. This usually increases the number of buyers and therefore price, all other things being equal.

Future "Volumes" of this Encyclopedia

If your elephant collectible(s) is left out, and you think it is an important, additional example of an existing type, or is an altogether new type (especially), please let me know. Send a high quality digital image (JPEG or TIFF), and related information about it (size, composition, year made, manufacturer or artist, value, your attribution preference) to my e-mail address. I will try to get it into the next volume of this ongoing "Encyclopedia."

Finally, although I believe the information presented to be accurate, if I have made an error in attributing an item, or in a description or value, I apologize. Please bring it to my attention via e-mail. After all, I want to learn as much about these collectibles as possible, but sometimes there just isn't enough knowledge out there to gather and present in a reasonable timeframe (I could have spent five years on this first volume!)

Section 2

Elephant Collecting

There are many facets to collecting anything, and elephants are no exception. For example, there are some interesting myths and stories related to elephant collecting. The most common is that one should only collect elephants with their trunk up; I don't personally believe it, and some collectors actually do the opposite. A myth borne of the Feng Shui craze is placing elephants near the entry of your home, facing in certain directions. The lore and myths of elephants and some aspects of collecting elephant-related things are captured *in An Enchantment of Elephants* by Emily Gwathmey, and: *Elephant Ancient and Modern* by F. C. Sillar and R. M. Meyler.

Why collect elephants in the first place? Well, for me, they are cool looking animals, their historical/ancestral forms are fascinating (e.g., anacus, mastodon, woolly mammoth etc.), and I believe they have been put on more things than any other animal. So there is a huge variety of things to collect. Just look at all the categories in Section 3; most people can find more than one category that interests them - independent of the elephant motif. So if you combine the interests, you can find years of enjoyment indulging those interests. Watch out though, for a pastime can turn into an obsession, something of which I can speak from personal experience; why else would I write this book! Also, a good reason to announce your interest in elephants is, when people know you collect elephants, you start getting them as gifts!

The Spousal Acceptance Factor: Managing Your Significant Other

Whether you are married or have a significant other, one thing is almost inevitable: conflict over your ever-burgeoning collection. Either in terms of size, amount of space taken in the home, or the financial angle, the spousal acceptance factor plays a part in your attempt to collect every cool elephant you see.

As your collection grows from the tens to the hundreds and then, for some, to the thousands, you need somewhere to put them. Depending on the size of your home, you first start out using available/existing space: in the curio with the dinner plates, on bookshelves along with Twain and Tolkien, and on what were, ostensibly, plant shelves. Then you need a dedicated space, because scattering them all over is messy and some are lonely, etc. So you either rearrange things to put them all on one set of shelves or in one curio, or you go out and buy or make dedicated curios or shelving.

Then the ultimate: you convert a room of your house, then your whole house, then buy or rent a building to display the elephants. That is exactly what some people do, as you can see in the following image of The Elephant Castle and Museum.

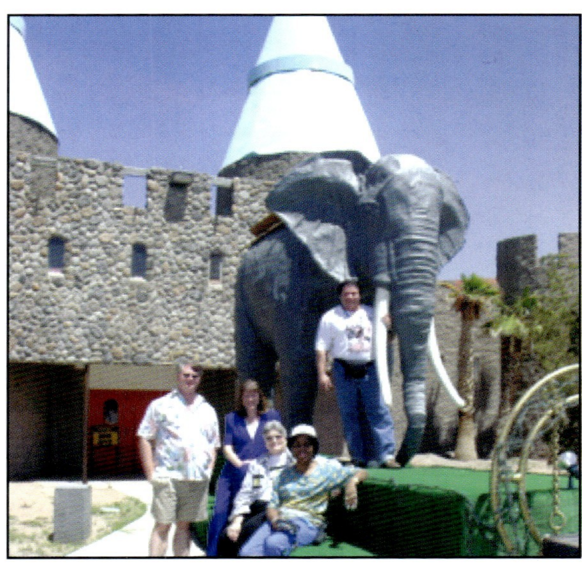

This is an image of me (nearest the elephant, which is fiberglass and life-sized!) and relatives in front of the Elephant Castle and Museum. It was located, until mid 2001, in Henderson, Nevada. It was run by Mitch Brown (email at Elephldy@aol.com), a wonderful lady who has been collecting for many years and has a wealth of information (and elephants) to share. Mitch also publishes the *Elephant Castle Gazette*, a newsletter for collectors, and runs an Elephant Collector's Club in the Las Vegas area. She plans to open a new museum in the Las Vegas area. I had a chance to visit the museum in May, 2001. It housed about 6,000 elephants of every kind imaginable. The building also housed a collectibles shop featuring, not surprisingly, elephants!

This scenario is fine if your significant other accepts or, better yet, joins you in your obsession. But if your relationship is not on solid ground, yielding ever more space and funds to your hobby can make them become resentful and angry. If they are not on board and amused and accepting of your hobby, NEVER buy them an elephant gift for their birthday or holiday! They will know for whom you really bought it!

How the Internet Is Changing Collecting

Let's face it, the Internet has changed just about everything, and collecting is no exception. For me, I reached an elephant collector's epiphany of sorts, when I first logged onto ebay and searched on the word: "elephant."

About 3,000 elephants came up for sale for one week! Now, at the time of this book's first printing, the number is about 10,000 per week, some being repeats of course. The point being, a great variety of elephant collectibles is available to collectors in an open market. It would take me the rest of my life, if then, to travel to all the places and shops, let alone individuals, to see all those elephants for sale. But on ebay they are all in one place. And that is just ebay; there are hundreds, perhaps thousands of other sites that have a goodly amount of elephanteria to look at. Refer to Section 4 for but a fraction of them that I have come across, and that helped me produce this book.

What to Collect: Specialization

If you have collected elephants for long, you probably realize that there are a lot of them out there!! Tens, if not hundreds of examples/instances in each of the categories that are listed here. (There are certainly some categories I did not include.) So, that means there are thousands of different basic types of elephants. That does not even consider the lower-level variations in, for example, color or size of a particular model. There are so many elephants, it is doubtful that anyone could collect every one, even with unlimited funds, no one can get every manufacturer or artist, every material, in every color and variety and size.

So what do you do? Specialize! Some have decided that only elephant figurines are elephant collectibles, and there are plenty of figurines to go around! Some may like tobacco-related items and so combine that with a love of elephant things, collecting elephant ashtrays, humidors, dispensers, matchboxes, etc. You can also divert an existing mainstream hobby, like numismatics or philately, to the elephant world, as there are plenty of examples of elephant coins and stamps. Another option is to collect elephants made on your birthday, or made during a certain era (e.g., Art Deco), by a particular manufacturer, or from a specific material. Or, be a "type" collector, wherein you try and get at least one excellent example of an elephant in each category. Another neat way to specialize is to collect one pachy from each place you visit. Or focus on elephants from the place you grew up. For me that would be Ohio, so whenever there is an elephant that relates to Ohio, I try to pounce on it!

Condition

In all areas of collecting, not just elephant collecting, one factor is of utmost importance: condition! It cuts across all categories of elephants - the better the condition, the rarer it is (as compared to used and damaged versions of the same thing), and the more it will appreciate, because other instances will become used/damaged over time. Therefore, ultimately, if the elephant is in the best possible condition, you will pay more for it.

So, if you can afford it, buy mint or near-mint items with little damage. That is, unless you find an unusual item or one so rare that affordability in any kind of future time frame would be out of the question. Not only does buying undamaged elephants pay off in case you ever sell, it will help preserve your piece of mind. I mean, you don't want to look through your collection and be reminded of that crack or chip or tear every time, right?

Now, that being said, there is nothing wrong with some normal wear (as opposed to "tear"). For example, if you buy a bronze that has been painted and is, say, a true antique (100+ years old or so), it is safe to say it is OK for there to be some minor paint problems - either small flakes or chips, or a rubbing/dulling of gloss. But not too much!! Or, if you buy an old magazine advertisement that has a minor margin tear that will "mat out," that seems OK too. Especially if you feel you will not get an opportunity to see/buy that exact item again, and it really appeals to you otherwise.

Of course you could take a purist stance and only look for perfect specimens. This is fine too, but keep in mind that it will take a lot longer to find specimens in perfect shape, and will cost more, likely much more, for certain items.

One example in my own case concerns an elephant coin minted in the 17th century that I wanted. In Very-Fine to Extra-Fine condition it runs around $750-1000. In MS-60 (Mint State-60) condition it is around $2000, in MS-63 it is about $3500 and in MS-65 (the highest I have seen for that coin) it is around $6000. Now, I would have settled for an Extra Fine (EF-40) considering my budget. But I have been a numismatist (coin collector) for 30 years, and I know that in that field condition is king! So I waited for at least an MS60, and actually found "one of the finest known" MS-63s for about $3500. In that case, paying the extra money for near-perfect condition for such an old coin was worth it.

This extreme attention to condition probably applies to few other types of elephant things (perhaps stamps, Steiffs, etc.), unless you happen to focus on that type. For example, if you specialize in elephant bronzes to the exclusion of all else, you can afford to be picky. Your time and money is focused and you want to get the finest possible examples within a particular category of elephants.

Things to be wary or aware of include:

- ivory vs. bone vs. synthetic - the hot pin test
- Bakelite vs. other plastics - the burning smell test
- post-ban ivory imported into US
- fake signatures, e.g. on Lalique
- reproductions and reintroduced models/names

As with all other collectibles, your familiarity with the subject will help you identify a reproduction from the real thing. Reading books, like those referenced in Section 4, monitoring internet auction sites like ebay, attending shows, flea markets, and live auctions, and talking to specialists and other collectors, all contribute to your knowledge and expertise.

Insurance

Heaven forbid something bad happens to your elephant collection. If you don't have them stored away in a safe place, like Fort Knox, a bank's safety deposit box, or an in-home vault (see Storage/Protection subsection), when disaster struck you would want some way of recouping the loss.

If your elephant collection starts burgeoning in terms of sheer numbers, cost/replacement value, or just sentimental value, you should consider getting insurance to cover them. Check with the insurance agent for the company that covers your home; many times the coverage for personal belongings is some percentage of the coverage for your house.

So if your house is covered for $100,000 say, and your personal belonging coverage/content is covered for 30% of the value of your house, then you are automatically covered for $30,000. Now, assume your furniture, TV/VCR/etc., clothes and kitchen wares etc., are worth $25,000, and your elephant collection is worth $3,000 (or you paid that amount over the years), then you may be covered to the extent you need to be. But, if in the same situation, you paid $25,000 for your elephants over 20 years, or they are currently valued at $25,000, you definitely want to add an insurance rider to make up the difference in coverage.

Storage and Protection

If you love your elephants like I do, and especially if you have a fair investment in them, you want to ensure you do what you can to prevent them from becoming damaged, or aging ungracefully. Collectibles are like anything else; the ravages of nature take their toll, eventually. Other than preventing the accidental break by storing your elephants out of reach of small children and animals, there are several considerations to protecting your elephants, including the following:

Humidity-controlled environment/curios. If you happen to live in Arizona, as yours truly does, this is not usually a problem, as the yearly swings are within tolerable levels (say 10% to 50%). But living near the beach or in the Midwest where swings of humidity can range from 5% in winter to 90%, is definitely going to play havoc with your collection. The only solution is a temperature and humidity-controlled environment. This can be achieved on a whole-house basis, on a room basis, or a smaller compartment like sealed tubes and plastic envelopes. The extent of the controlled environment is a function of your budget and needs. To mitigate the effects of humidity on wood for example, use lemon oil or tung oil on the surface and get it into any visible cracks; this will help seal and moisturize the surface.

UV protection. Keep your collectibles, especially plastics and ephemera, paintings and prints etc., away from UV sources, especially the sun shining through the unfilmed/untinted window, and strong light sources. Prints, paintings, posters and other displayed ephemera can be framed with UV-safe glass or plastic.

Dirty fingers. Your body produces oils. If you touch your computer or TV screen you can see an immediate smudge, so what do you think that does to paper or other materials? After a few years, you will see that smudge on your elephant print or your brass elephant as it collects dirt from the air. So use stamp tongs and wash your hands thoroughly with soap just before handling your elephants. Of course, you can also wear gloves. Clean cotton gloves are inexpensive, or you can get inexpensive disposable polymer film gloves, or reusable gloves like Polygenex Nylon Lab Gloves.

Bugs. Bugs, like woodborers and paper beetles can destroy a great collectible in no time. Make sure you inspect your wood and paper-based elephants regularly and spray the collection area/room with something like FICAM at appropriate intervals. Be sure to quarantine a wood elephant that came from a foreign country for a while, up to a month or so, to ensure that it contains no pests like borer beetles.

Air. Environmental gases result from the burning of fossil fuels, auto emissions, decay of plants, ocean surfaces and even outgassing from paper products and certain plastics. Sulfur and chlorine are especially damaging to metals. Protection options here range from conservation mounts/plastics, and acid-free rag-based mat board and backing board for mounting prints and advertisements, to zippered bags treated with chemicals that intercept any damaging gases and trap them or neutralize them before they hit your collectable. In fact, one brand name of such a zippered bag is called "Corrosion Intercept" available from University Products Inc.

To protect ephemera do as philatelists have done for years; insert your ads, stamps, magazines, prints and labels (that you are not going to otherwise frame) into archival-quality sleeves and mounts. Stay away from anything labeled PVC, as that has plasticizers than can bleed into your art. Polypropylene, polyethylene, and Mylar polyester are safe alternatives. Do not hinge your stamps, or anything else! My favorite sleeve/album combination is made by SuperSafe. Vario makes great Mylar polyester archival-quality mounting pages for different size stamps, but these are also ideal for trade cards, labels and other small, flat artifacts. I also use Larson-Juhl Conservation Corner Mounts for mounting prints and posters and ads onto acid-free mat boards that are to be framed (without having to actually "mat" each piece).

Coins, like stamps and ephemera, can be protected and displayed using any of a number of enclosures. Non-PVC (e.g., Mylar) coin flips are inexpensive, and your coins can be displayed individually. Corrosion Intercept also makes coin pouches that add a further element of protection to coins. If you buy a PCGS or similarly graded coin, it will come in a single plastic, sealed, airtight mounting package great for display. Multi-coin plastic displays in several size configurations, including mixed sizes are handy when you have several very different coin sizes but want to display them all together.

Information related to storing and protecting your particular collectible can be gained from resources such as: librarians, framing shops, art shop owners, museum curators, and books on conservation techniques including: *Guide to Environmental Protection of Collections,* Barbara Appelbaum, 1991 and *Conservation Concerns: A Guide for Collectors and Curators,* ed. Konstanze Bachmann, 1992.

Fixing Broken Elephants

Of course, the old adage: "an ounce of prevention is worth a pound of cure." applies to elephant collectibles as well. But the sad day inevitably comes when, by moving an elephant in the home, or by shipping accident or other mishap, an elephant becomes damaged. Some collectors buy damaged elephants and either fix them or leave them alone, claiming it increases the charm or "character" of the find. Others buy a damaged elephant if it is especially rare or if a perfect one would be too expensive.

One note of utmost importance: it is not advisable to clean or otherwise repair true antiques, unless a professional does it and you understand the consequences. For some antiques, refinishing, restoring, or repairing certain flaws actually diminishes value, even though it may make the elephant look better cosmetically. Some dirt or dust can be removed on pottery, metal, or wood with nothing more than a damp cloth.

The most common damage I have seen is missing or broken tusks. For many types of elephants - tusks can be repaired or replaced. Wood, plastic or ivory tusks can be re-created with a little ingenuity and skill. For example, a dowel rod of the appropriate length and diameter, soaked in water or put in a steamer for a few hours, can be bent into the appropriate curve and held there for several hours to set the shape. Then, further shaping with carving tools can produce a most pleasing replacement wood tusk. Add paint or stain to match the elephant or an existing tusk. Replacement ivory can be carved from mammoth ivory (legal and available) to replace tusks and toe-inserts or missing pieces on ivory elephants.

For common pottery elephants, breaks or cracks can be repaired at home using common glue or epoxy cement. More expensive elephants can be taken to repair shops that specialize in such repairs. They usually re-break the piece, treat the surfaces, re-glue and then, the most important step, re-glaze/fire the piece. The result is a repair than only trained professionals with a microscope could find. Highly recommended for that favorite, expensive piece.

Metal elephants can be repaired by skilled metalworking artists. I have an old brass elephant box that had missing tusks. I took it to a metal artist who used brass rod to create and re-solder the tusks into the holes. Natural aging should even out the patina. Brass, bronze, aluminum, copper, and chrome elephants can be cleaned and protected with the common, non-abrasive metal cleaners and polishers. Rubin-Brite is a museum-quality cleaner/polisher that leaves a carnuba-wax protective finish on the metal. Iron and steel elephants can rust, which requires more work. A rust remover gel, followed by 0000 steel wool cures most rust spots. Again, for older, rare or true-antique metal elephants, unless the corrosion is so advanced or bad that it further endangers the elephant, leave minor discoloring and surface blemishes alone.

Ephemera, paper images, prints, posters and paintings, can be repaired by professionals if the item is pricey or rare, and some repairs can be done by the home hobbyist. Pencil marks on paper can be removed by gently rubbing with an eraser-like material called "Magic-rub" by Sanford. A more thorough cleaning can be gained using Lineco's Document Cleaning Powder. Paper items can be de-acidified using Bookkeeper Deacidification Spray. Tears can be repaired using Lineco's transparent mending tissue.

Lastly, a great reference on caring for your elephants (or any collectible) is: *Kovels' Quick Tips: 799 Helpful Hints on How to Care for Your Collectibles* (Kovel's 1995).

Sources of Elephants

Elephants can be found almost anywhere other products are found. Because there are so many types of elephants, even specialty stores (like a kitchen & bath shop) may have that obscure elephant needed for your collection. Here are some places I have found elephants:

- Almost any retail store like Wal-Marts, Hallmark, Sears has elephants - mostly mass-produced
- Estate Auctions
- On-line Internet auctions like ebay.com, amazon.com
- On-line antique stores and malls like www.rubylane.com
- Antique stores
- Flea markets
- Yard Sales/Garage Sales
- Looking for elephants wherever you go on vacation.

My Favorite Types of Elephant Collectibles

My favorite categories are:

- Older and/or rarer figurines of certain manufacturers/artists (e.g., Meissen, Teplitz-Amphora, Daum, Loetz, Royal Doulton, Delft, Wedgwood)
- Minerals
- Ephemera, especially older advertisements with elephants, die-cuts, and trade cards
- Coins and Stamps
- Inkwells

- Incense Burners
- Bookends
- Cigar cutters.

My want list includes the following "Elephants To Die For" (E2D4). The ones in the list, that I have actually shown in this book, have an (E2D4) after their caption in Section 3.

Amphora/Teplitz Region
Armani
Baccarat 1993
Bakelite
Barovier Glass/Seguso
Barye Running, original factory
Bernard K. Passman black coral
Brazzini Nuts Tin
Burmese Jade
Carter Vanishing Elephant Poster
Ceasar Denarius XF or better
Ceylon 1800s stivers coin
Chase Bookends
Chinese Roof Tile Man on elephant
Collas Cafe original poster
D. Riggs (several)
Daum Pate DeVerrve Art Deco Glass (3 sizes)
Dedham Plate/Bowl
Delft Tile 16/1700s
Disney Dumbo Cell original
Eagle Cigarettes Porcelain Sign
Elephant Brand Coffee Tin
Elephant Oil Can
Faberge Precious/semi-precious stone carving
G. Price (Large)
Ganesha-old bronze & sandstone
Goebel_Hummel Nativity
Hagenauer
Herend, Large
Herend, Small on Ball, Black Fishnet
Ivory Bridge & Scrimshaw & Large carving
Jack Bryant large detailed bronze
Kay Finch
Lalique Java
Lapis Large
Lenci
Lladro: several more!
Loet Vanderveen (several!)
Malachite Large
Matchsafe brass – Victorian
Meerschaum Pipe
Meiji-period (Japanese) Bronze
Meissen (several!)
Neiman
Nymphenburg
Ronson Bronze Striker Lighter
Rookwood - 2444C/D trunk down in Blue and also trunk up
Royal Doulton Flambes, Large Fighting elephant
Staffordshire figurine & spill vase
Steiff - 1900s
Steiff - 1955 75th Anniversary
Steuben
TWA To India 1946 Ad/TWA Poster
Van Briggle (on base and separate)
Vienna bronze figurine
Vienna bronze striker lighter in cloak
Zanesville

Section 3

Pictorial Encyclopedia of Elephants

This is the section you have been waiting for: 1000 images of all kinds of elephant collectibles sorted into 140 categories, one subsection per category. In the beginning of each subsection are a few words about the category. As mentioned in Section 1, many of these categories could be combined into others, or some images in one category could be split out into two or more separate, finer-grained categories. But this is how I did it and I hope you can easily find what you are looking for.

The format for this section is a series of alphabetically-sequenced categories, within which are a set of images with the following information (if I was able to find the information!): a brief description, size, year(s) made (or approximation), a valuation (range - see disclaimer in Section 1), and an attribution (if the owner wanted to be mentioned). If the elephant is on my Elephants To Die For list then (E2D4) is at the end of the caption.

Section 03.001 Advertisements

This subsection presents images of one of my favorite categories: paper advertisements with elephants in them. Elephants have been used since the 1800s to advertise virtually everything. The most common products sold using elephants, include: alcoholic drinks, automobiles, gasoline and oil, and recently, computer-related products like memory. In compiling these images (which are a fraction of those I have in my own collection), I concentrated on those wherein the elephant is a central part or theme composing the message of the ad, power or memory prowess for example. Additional advertisements using elephants are in the "Posters" section and in the "Advertisements and Commemorative Figurines and Sculptures" section.

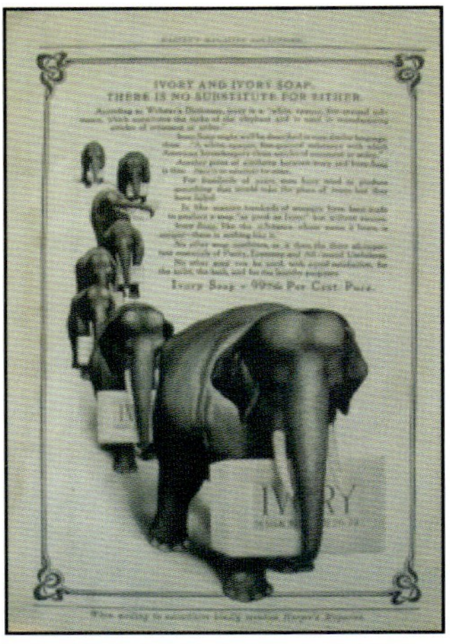

Ivory Soap ad from Harpers Magazine. 10" H x 6" L. 1907. $10-15. *From the author's collection.*

The Trade-mark Ethyl gasoline ad 13.5" H x 11" L. 1948. $6-10. *From the author's collection.*

Jello ad. 14" H x 10.5" L. c. 1960s. $6. *From the author's collection.*

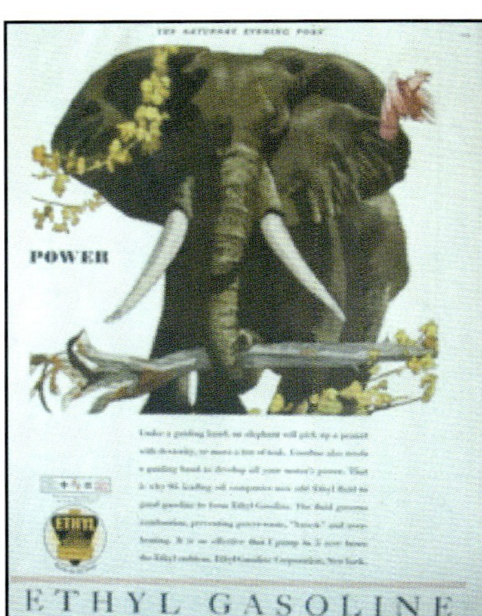

Ethyl Gas ad, Saturday Evening Post. 13.5" H x 10.5" L. 1931. $10. *From the author's collection.*

Johnson Printing Inks ad, Charles E. Johnson and Co. 12" H x 8" L. 1928. $5-10. *From the author's collection.*

Honeywell ad 13.5" H x 10" L. June 1966. $7-10. *From the author's collection.*

Imperial Hiram Walker's Whiskey ad 13.5" H x 10.5" L. 1942. $7-10. *From the author's collection.*

Anderson Consulting ad that appeared in several print locations and sizes. 10.5" H x 9" L. 1999. $3-5. *From the author's collection.*

Section 03.002 Ashtrays

This subsection presents images of ashtrays. Other ashtrays, that are also political commemoratives (e.g., GOP-related ashtrays), are in the Political/GOP section. Bronze, pottery and quartz-based ashtrays seem to be the most popular materials.

4-piece ceramic ash tray set, Occupied Japan. 3.5" H x 6" L. 1950. $30. *From the author's collection.*

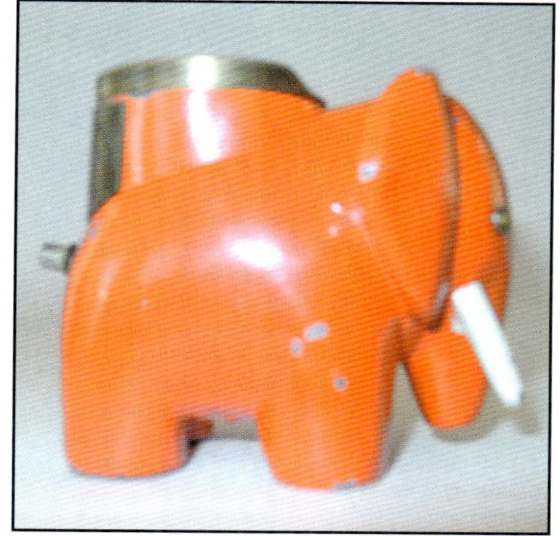

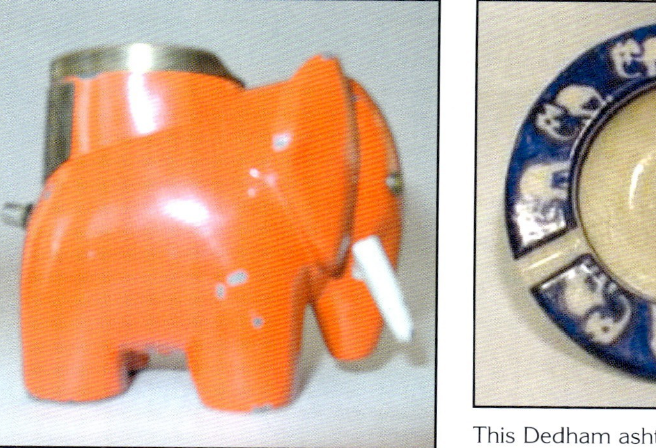

Painted metal figurine with brass eyes, whose hollow body functions as an ashtray. Push a brass button near the tail and the brass ashtray on top of the elephant opens to the body of the elephant. Made by Brown and Bigelow of St. Paul, Minn. Brass plate on rear reads: "Compliments of B. E. Thoele Dental Laboratories 'The Laboratories of Personal Service' St. Cloud, Minn." 3.5" H x 3.5" L. c.1950s. $25. *From the author's collection.*

This Dedham ashtray has the familiar crackle glaze and the circle of blue elephants (no baby elephant on this one). c.1900. $300-400.

Hamilton chrome-plate ashtray, inscribed Joseph Goder Incinerators. 1.5" H x 4.25" L. c.1950s. $15-25. *From the author's collection.*

Depression glass ashtray made by Greenburg Glass. 3.5" H x 6" diameter. 1920s-30s. $35-75. *Courtesy of DJPowers.*

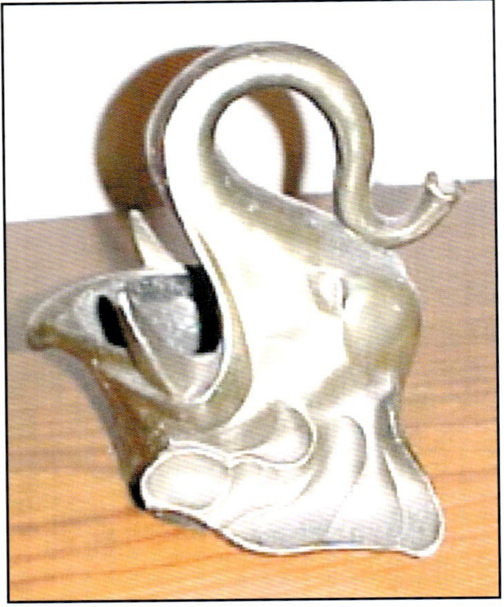

Frankart, a New York firm made this Art Deco metal (white lead composition, spray painted) ashtray. It is in the form of an elephant head with raised trunk; it is signed behind the left ear. These have a metal insert and felt on the bottom. 7" H x 6" L. c.1920s-30s. $100-150.

A green ceramic ashtray advertising Fremlins Elephant Ale by the Fremlins Brewery. 1950s-60s. $50-75. *Courtesy of Mary Marwick.*

An ashtray by artist Sascha Brastoff. 1953-73. $25-100.

This bronze ashtray features a little elephant standing on the side. Marked: "L.V. ARONSON 1924" on the bottom. 1924. $150-200. *Courtesy of Scott Moore.*

Stangl pink elephant ashtray made for the New Year by Kay Hacket. This piece is listed in *Duke, 121 n 3915*; not many were produced. 9" H x 9" L. c.1950-60s. $150-225. *Courtesy of Michael Locati.*

Section 03.003 Badges and Patches

This subsection presents images of badges and patches given to award persons for valor or to commemorate a milestone in a person's career, or for belonging to a particular group. One example would be the Oakland Athletics baseball club, whose mascot is the elephant; an image of an elephant is found on their jerseys, patches, decals, flags and banners. Also see the Trademarks section for a black powder patch with an elephant as trademark.

A State Riot Police badge from Bremen, Germany; this elephant is the insignia for Sonder Einsatzzug 144 (the unit name). 5" H x 4.5" L. c.1990s. $10. *From the author's collection.*

Patch from the Thailand border police, a quasi-military unit originally funded by the CIA to serve as a counter-insurgency group in the early 1970s. They are still active along the borders and work with the U.S.'s DEA in anti-narcotics operations. 3" H x 2.75" L. 1970s. $5-10. *From the author's collection.*

Patch from the Kelly Air Force Base (U.S.) Logistics Support Squadron in Texas, U.S.A. 1990s. $5-10. *From the author's collection.*

White metal cap badge of Queen Alexandra's own Royal Hussars (2nd pattern), probably made by Firmin and Sons. 1915-1928. $30+. *Courtesy of John Whitworth.*

Section 03.004 Bags

This subsection presents images of both paper and cloth bags that have an elephant on them as a company logo or advertising something associated with elephants, like peanuts.

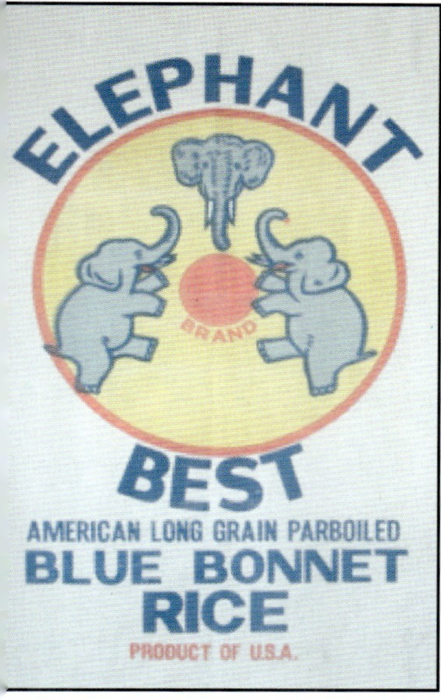

Panel from Elephant Brand Best Blue Bonnet Rice bag: 38" H x 23" L. $5. *From the author's collection.*

Jumbo Bemis paper sack for 12 pounds of Jumbo Flour made by the Rocky Hill Milling Co. 19.5" H x 11" L. c.1940s. $10. *From the author's collection.*

Big Jo flour sack held 98 pounds of flour. 38". 1927-1931. $250. *Courtesy of Doug and Sandy Feiner and Roger and Beverly Koch.*

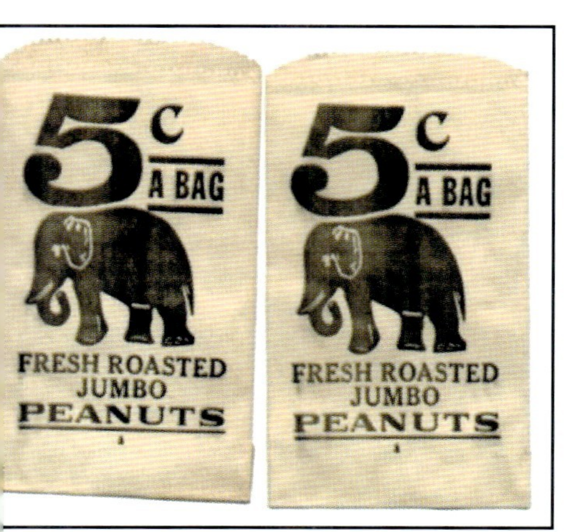

Jumbo Peanuts paper sacks. 1930s. $6-10. *Courtesy of Mary Minges of Minges Antiques.*

The Three Ring Coffee Bag has red background color with black skirted elephant balancing three rings on her nose (snout). It's an unused one-pound bag made by Kitchen Products, Inc. Distributors Chicago, Illinois. 10.25" H x 3.25" L. c.1950s. $3. *From the author's collection.*

Section 03.005 Banks

This subsection presents images of elephant banks. Many collectible still and mechanical elephant banks were made of cast iron from 1900 to about 1940, by companies such as Griffith, Hubley Manufacturing, and A.C. Williams.

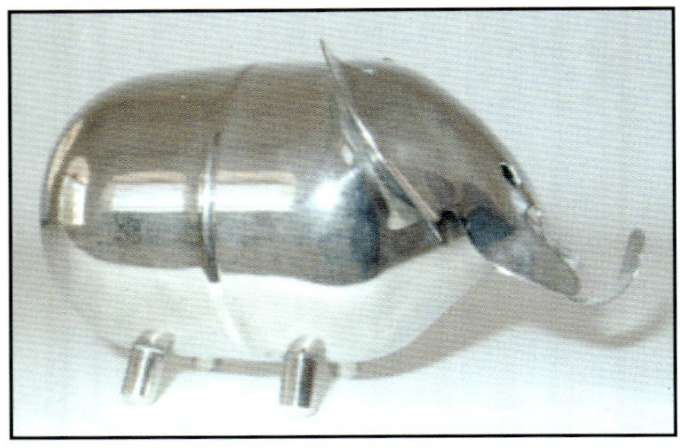

Tin/metal, Lunt bank. 3" H x 5.5" L. $15-25. *From the author's collection.*

Copper bank from the Hang Seng Bank Ltd. 5.5" H x 7" L. $25. *From the author's collection.*

Porcelain Enesco bank, Made in China. 5" H x 8" L. 1979. $10-15. *From the author's collection.*

Elephant with howdah, cast iron bank made by A.C. Williams, U.S.; see *Moore 2000, 84 n 474*. 4.875" H x 6.375" L. 1910-1930s. $50-75. *From the author's collection.*

A hand-painted reproduction of a vintage iron mechanical bank, made in China. Originals made by Hubley, for example, can range to several hundred dollars. 4.75" H x 6.5" L. c.1990s. $15. *From the author's collection.*

J. Chein tin bank. 5.5" H x 5" L. 1930s-40s. $100-120. *Courtesy of ebay user "Thirsty1."*

Ceramic bank from EDAS in Brazil. 5" H x 6" L. c.1980s. $15. *From the author's collection.*

Hubley cast bank. Shown in *Moore 2000, n 462*. 3.875" H. c.1920-40s. $300-500. *Courtesy of Dan of Adamstown Antique Gallery.*

The Grapette Bottling Co. in Camden, Arkansas, produced a soft drink syrup which they distributed in glass figural banks such as this elephant, starting in 1947. This one is a rare "slick eared" elephant bank, only used for a very short period of time due to it being rejected by the public and continually breaking in the filling machines. It is extremely rare and highly sought after by bank and soda collectors. This bank has the original shipping label attached from the Owens-Illinois Glass Company in Toledo, Ohio. The shipping tag is dated 9-4-52 and is marked "new merchandise." 7" H x 3" L x 2.5" D. 1952. $350. The common Grapette elephant bottle bank is worth about $10-20.

Cast sitting elephant bank made by Vanio. 5" H x 3" L. 1936. $50-100. *Courtesy of Stan Alford.*

A cast metal, advertising elephant bank, finished in a bronze wash color. The sides of the base are inscribed: "IVORY SALT—WORCESTER SALT—IODIZED SALT—WORCESTER SALT." Also found on the base is the door from which you could extract your savings. Next to it is the slot in which you would slip your coins or dollars for saving. Inscribed on the bottom is "BAUDIS METALCRAFT COL. NEW YORK." 4.25" H x 2" L x 3" D. 1910-30s. $35-50.

Section 03.006 Baskets

This subsection presents images of baskets made, usually, of wicker. Although wicker baskets may be among the most inexpensive elephant collectibles, I have yet to find a collector who doesn't have at least one. And baskets are functional too, usually holding anything from planters and dried flower arrangements to magazines.

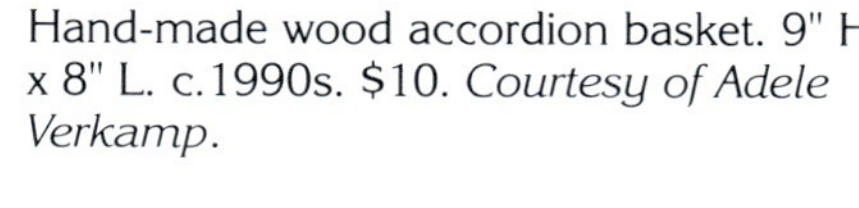

Hand-made wood accordion basket. 9" H x 8" L. c.1990s. $10. *Courtesy of Adele Verkamp.*

Wicker basket. 9" H x 17" L. c.1980s. $20. *From the author's collection.*

Section 03.007 Beer Cans and Bottles

This subsection presents images of beer cans and bottles with an elephant on them. Since several beer manufacturers produce an "elephant" brand, you will find other items that sport or advertise the beer using elephants: glasses, signs, coasters etc. In addition to those shown, I know of other cans and bottles sporting elephant images including Delavan 1986 Beer and Halida Beer from Vietnam.

Brown glass beer bottle in 1-quart size by Pfeiffer Brewing Company of Detroit, Michigan. The label was created by Johnnie Pfeiffer. 10" H x 3.75" L. 1948. $25-40. *From the author's collection.*

Imported Elephant green glass beer bottle by Labatt; USA branch based in Norwalk, Connecticut. 9" H x 2.25" L. 1990s. $3-5. *From the author's collection.*

Rogue beer can from South African Breweries, LTDF, in Johannesburg, South Africa. 4.75" H x 2.5" L. $3-5. *From the author's collection.*

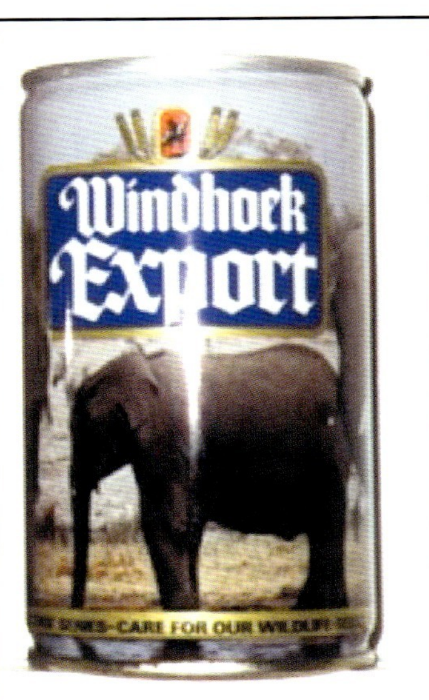

Elephant Red large beer can. 6.25" H x 3.25" L. 1990s. $5. *From the author's collection.*

Relatively rare Ginger Beer stoneware bottle. It has the Ginger Beer trademark of Jumbo Bottling Works of Cincinnati, Ohio on a 3-color label. 8 oz. size. c.1900-1910. $200-300. *Courtesy of Bill Champion.*

Beer can issued by the South West Breweries in South West Africa. 340 ml. size. 1970s-80s. $7-10. *Courtesy of Klasines Nijmeijer.*

Section 03.008 Beer Taps

This subsection presents images of beer taps, one with a prominent plastic or resin elephant atop it, the other with a woolly mammoth painted on it. I know of one other elephant beer tap; it is for the Lost Coast Brewery, in Eureka California, called Indica; the art is an abstract elephant king.

Rock Ice Amber Lager beer tap by Rolling Rock Brewery in Latrobe, Pennsylvania, blue background with mastodon or wooly mammoth image. 9.5" H x 5" L. 1990s. $20-30. *From the author's collection.*

Elephant Red beer tap with brown composition elephant. 13" H x 7" L. 1990s. $30-40. *From the author's collection.*

Section 03.009 Bells, Gongs, and Chimes

This subsection presents images of elephant bells, gongs and chimes. Most of these are made of ceramic, glass, or metal, owing to the need to make sound when struck.

Wood elephants holding gong. 14" H x 21" L. 1930s. $250. *Courtesy of Ted Capell.*

Brass elephant holding brass bell with wood ringer. 4.75" H x 6.5" L. 1980s. $30-40. *Courtesy of Adele Verkamp.*

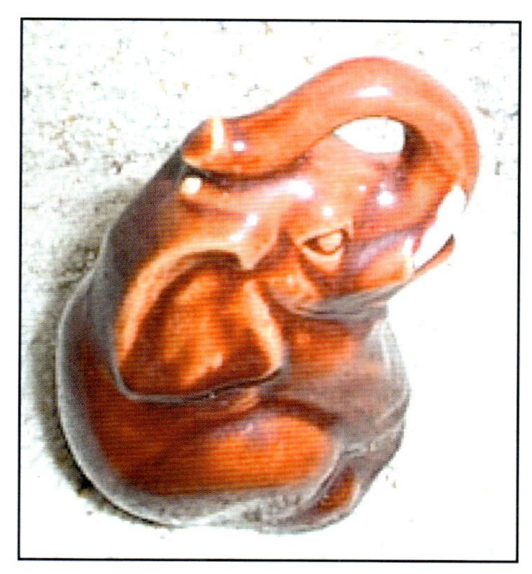

Rosemeade elephant bell. 4" H x 3" L. 1950s. $150-200. *Courtesy of Don Daley (a member of the North Dakota pottery collector's society).*

Ceramic bell from Ihsing, in China. 6" H x 4" L, $20. *Courtesy of Adele Verkamp.*

Brass bell. 3.5" H x 2" L. $5-10. *From the author's collection.*

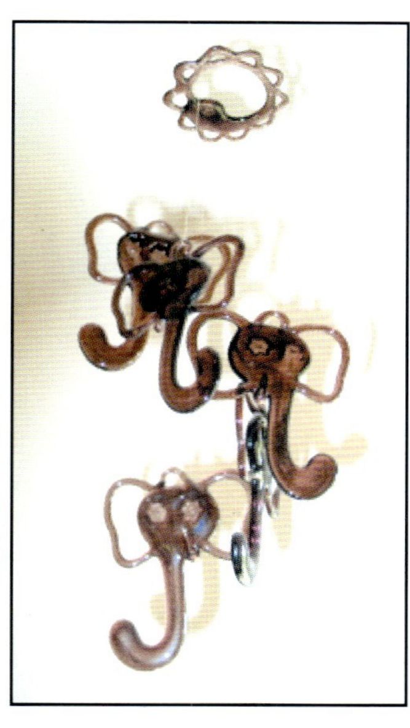

Purple glass wind chimes. 6" H x 4" L each. 1970s. $15. *From the author's collection.*

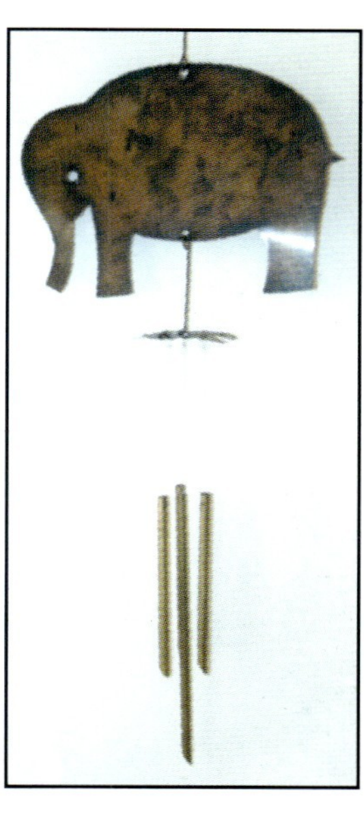

Brass and copper wind chimes. Elephant is 4" H x 8" L. 1970s. $15. *From the author's collection.*

Ceramic bell marked: "VC CTI, Japan;" part of a kitchenware series. 4.75" H x 3" L. 1970s. $10-15. *From the author's collection.*

Reed and Barton silver-plated hotel desk bell. Note that Reed and Barton's trademark is an elephant; see Trademark section for an example. 1885. $100-200. *Courtesy of Katie Babbitt.*

Section 03.010 Bookends

This subsection presents images of elephant bookends. This category has lots of specimens from many different manufacturers created in many different materials. Bookends are a popular collecting area in their own right, and an elephant collector could concentrate on just bookends and compile a very nice, valuable collection. Most of the best potteries (e.g., Rookwood, Cowan) and metal-products companies (e.g., Chase, L.V. Aronson, Hubley) produced now-classic bookend forms (at least to the elephant collector).

In *Bookend Revue,* Seecof mentions that the first bookends were really portable bookcases, made in approximately 1800 (even though books have been around since about 2nd century). Then, bookslides were made, essentially a single shelf or rail serving as a shelf, connected by what we would now call bookends. At the turn of the century, the first freestanding bookends were "invented" by Melville Dewey in U.S.; they were simple flat thin steel bent at 90-degrees, similar to those you can buy today. Bookend design motifs followed the art world's design trends: Arts and Crafts, 1861-1920; Art Nouveau, 1890-1910; and Art Deco 1910 – c.1930s.

Pair of bronze (possibly L.V. Aronson) bookends; see (Kuritzky 1997, 118). 5.5" H x 7.5" L. c. 1920s. $100-125. *From the author's collection.*

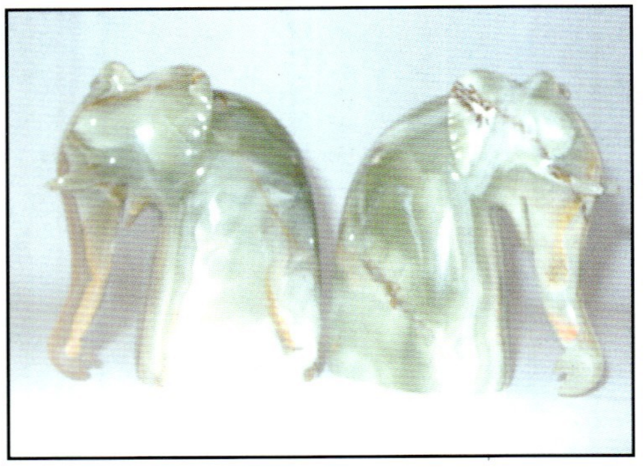

Pakastani carved soapstone bookends. 6.5" H x 5.75" L. 1990s. $40-50. *From the author's collection.*

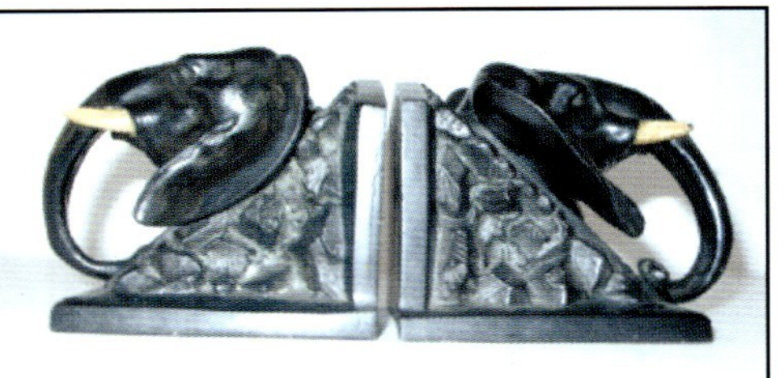

L.V. Aronson Elephant Head bronze bookends. 4" H x 5" L each. 1922. $100-125. *From the author's collection.*

Ronson (also known as LVA or L.V. Aronson) bronze bookends were featured in the 1925 Ronson All Metal Book Ends catalog and described as "Book Ends No. 10854, Elephant Design, All metal Juvenile book Ends I miniature size. Supplied in Royal Bronze." The front cover of the catalog describes Royal Bronze as: "A rich, lustrous, greenish brown, hand-relieved in bright green." Paper sticker signed: "Ronson All Metal Art Wares." Each piece is also signed on a separate metal plaque: L.V. Aronson 1923. See *Kuritzky 1997, plate 454*. 4.5" H x 3.75" L. 1923. $75-125. *From the author's collection.*

Lefton #H5456 porcelain bookends. 5" H x 4.75" L. 1960s. $25-35. *From the author's collection.*

Iron bookends. 4" H x 5.5" L. 1926. $75-125. *From the author's collection.*

Iron bookends by A. C. Williams Co. 3.5" H x 4" L. 1930s. $25-40. *Courtesy of Jeannie Delauter of Born Again Antiques.*

One-of-a-kind set of bookends that were made by the Armor Bronze Company on New York. They are quite detailed, solid bronze, and feature a turquoise accent paint. These bookends came with green felt bottoms and a company sticker. 6" H x 7" L. early 1900s. $400. *Courtesy of Jeanette Combs.*

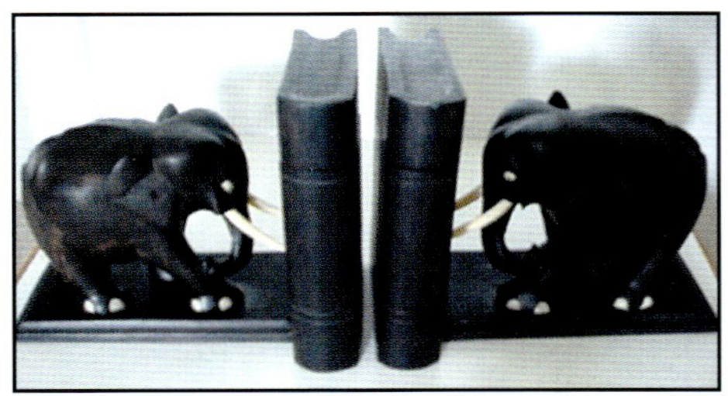

Ebony and ivory bookends. 6" H x 8" L. 1950s. $75. *Courtesy of Jadien Tokman.*

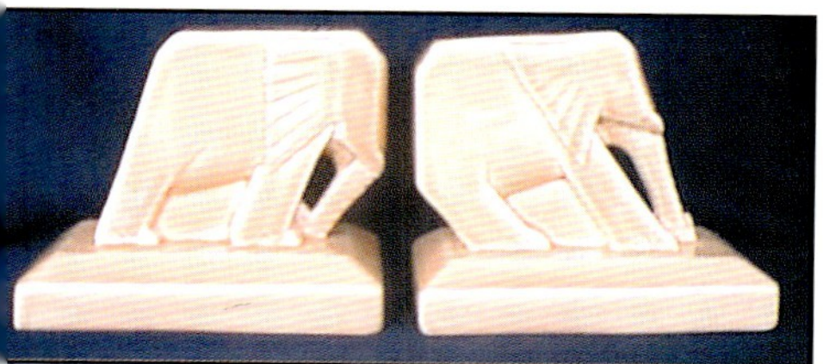

Extremely rare Cowan bookends gold-stained Art Deco; referred to as the "Push-me, Pull-me" elephants. Guy Cowan made pottery in Rocky River, Ohio, a suburb of Cleveland, from 1913 to 1931. A stylized mark with the word "Cowan" is on each base. There is an "E" stamped on each piece as well, defining the artist. 4.75" H. 1930s. $1200. *Courtesy of Paula Fulcher.*

Porcelain bookends of sitting elephants, made in Germany. 5" H. 1930s-40s. $100-150.

A pair of pottery bookends from Chester Nicodemus. 4" H x 5" L. 1940s-50s. Value $750. *Courtesy of Ronnie Sanford.*

Bronze bookends from Nuart in New York City. 6" H. 1931. $200. *Courtesy of Valerie Dianna.*

These Art Deco-style, detailed bronze bookends were made by the Pompeian Bronze Company and have the original felt and labels intact on the bottom. The embossed label reads: "Made by Master Craftsman, POMPEIAN BRONZE." Pompeian Bronze Company was located in Brooklyn, New York in the 1920s and 1930s. 6" H x 5.75" L. c. 1928. $300-400. *Courtesy of Devin de la Torre of spiritedelements.*

Ronson bronze bookends with elephant sitting up on book. 4.5" H. 1920s. $75-125. *Courtesy of turnit2cash.com.*

Rookwood porcelain bookends 1919 #2444D in white matte glaze. This bookend model was made in a variety of colors including Cerulean Blue matte in 1923 and an ivory matte in 1924. 5.25" H x 6" L. 1921. $300-400. *Courtesy of Collin Fleming. (E2D4).*

Set of trunk-up (unusual) Rookwood bookends. The colors are varying shades of blue, medium to a deep navy. They are marked with the usual Rookwood logo as well as shape #6124 and the Roman numerals for the date. 7.25" H x 5.5" L. 1937. $600. *Courtesy of Patti's Past Perfect Pottery*

This cast metal bookend is a large bull elephant that rests with one knee on a pillar of rock and its trunk curled back to its forehead. It was made by Sarsaparilla Design, a metal foundry in West New York, New Jersey, that recently went out of business after over 50 years of production. 7" H x 5.5" L x 5" D. c. 1980. $100. *Courtesy of Richard Jay Silverman.*

Carved tiger-eye wood and ivory bookends with remarkable grain. 8.25" H x 9.5" L x 4.25" D. $250. *Courtesy of Randy Esada Designs Inc., Los Angeles, California.*

Section 03.011 Bookmarks

This subsection presents images of bookmarks. I have seen several other styles and materials including ivory, leather, various metals, and printed images on plastic-coated paper or cardboard.

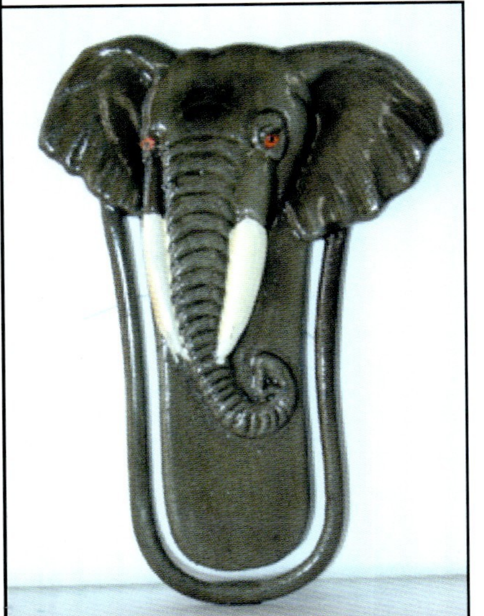

Painted wood bookmark from China. 3.75" H x 2.5" L. 1990s. $5. *Courtesy of Adele Verkamp.*

Brass and enamel bookmark. 6" H x 1.75" L. $5. *From the author's collection.*

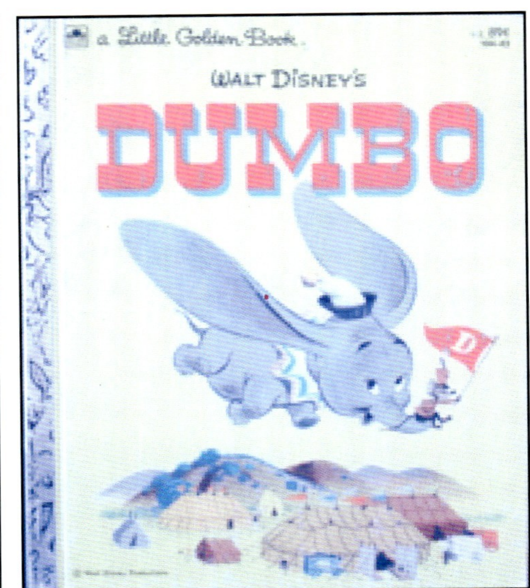

Walt Disney's Dumbo, a Little Golden Book. 1947, 1939 (1st Ed). $10. *From the author's collection.*

The Elephants Child, by Rudyard Kipling, part of the "Just So Series" published by Garden City. The elephants were illustrated by F. Rojankovsky. 1942; the 1900 first edition was published by Curtis Publishing. $5-10. *From the author's collection.*

Section 03.012 Books

This subsection presents images of books about elephants, real and imaginary. There are hundreds (maybe thousands) of elephant books, and I know of one collector, Wayne Hepburn, who focuses on elephant books. With his help, I have compiled a list of hundreds of books, all the elephant books I could find. I did not include it in this volume because it takes up so much room, and there are no images for the vast majority. But I can supply this list; email me for details.

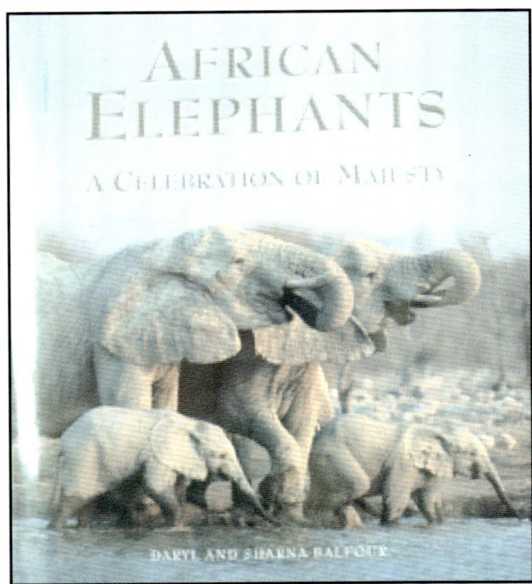

African Elephants, by Daryl and Sharna Balfour, Abbeville Press, Inc. 1998. $35. *From the author's collection.*

Oliver, by Syd Hoff, HarperCollins Children's Books. 1960. $10 *From the author's collection.*

Modoc: The True Story of the Greatest Elephant That Ever Lived, by Ralph Helfer, published by Harper Collins. This book tells the story of an elephant born in Germany in 1896. 1997. $10. *From the author's collection.*

Babar and His Children, by Jean De Brunhoff. 1996; the hardcover was first published by Random House in 1938. $5-10. *From the author's collection.*

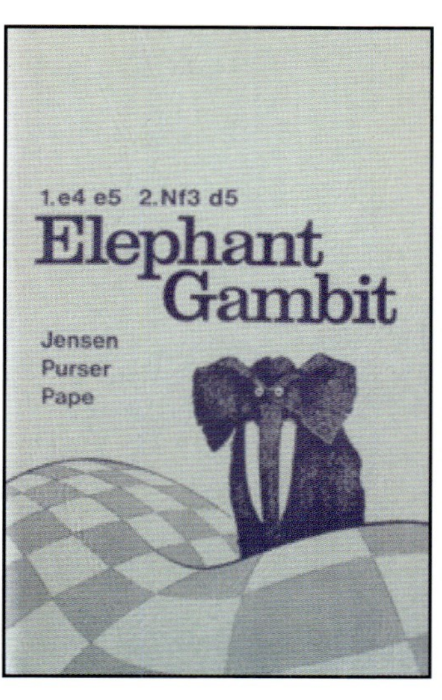

Elephant Gambit, by Jensen Purser Pape, soft cover, published by Blackman Press. It explains the chess moves referred to as the "elephant gambit." 8.75" H x 5.5" L. 1988. $5-10. *From the author's collection.*

This is one of the only other books dedicated to elephant collectibles. Titled *An Enchantment of Elephants*, by Emily Gwathmey, it was published by Clarkson/ Potter Publishing of New York in 1993. It is out of print now. Emily specializes in ephemera and is a deltiologist (collector of postcards), therefore, many of the images in her book are in those categories. 9" H x 9" L. 1993. $25. *From the author's collection.*

Section 03.013 Bottle Openers

This subsection presents images of some elephant bottle openers.

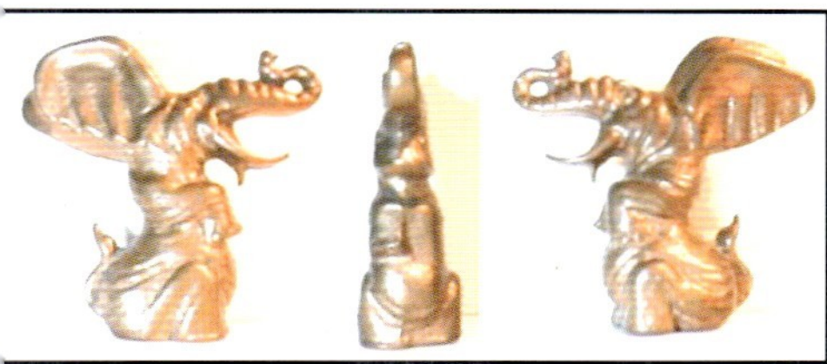

Brass opener. 4" H x 3" L. 1960s. $50.

Hangenauer brass bottle opener model #213 made in Austria. c. 1920s. $50-75. *Courtesy of Joseph Cantara.*

This heavy brass bottle opener is a variation of the model #F 48 (made by Wilton Products in the 1950s, value: $30) listed *in Identification Guide for Figural Bottle Openers* published by the Figural Bottle Opener Collectors (FBOC) Club. This opener is from Riverside Brass in Canada. There are no markings. 3.375" H. c. 1970s. $15. Some of these were made with GOP impressed on them. They date from the 1920s, are usually pink, and valued at $50-70. There is also a similar opener made by Hamilton Foundry. *Courtesy of Barry*

Section 03.014 Bottle Stoppers

This subsection presents images of elephant bottle stoppers. Anri is perhaps the most well known maker among collectors, making woodcarvings since 1912 in Italy.

Brass and pewter Italian bottle cork with neck chain. Has letters C (copyright) WTU on elephant's base. 3" H x 2" L. $10-15. *From the author's collection.*

Rare Anri wood articulated bottle stopper. 3" H x 2" W. 1940s-50s. $225. *Courtesy of Dick.*

Brass stopper with crystal eyes by Hans Turnwald Sign of Design. 3.5" H x 2.25" L. 1990s. $10-15. *From the author's collection.*

Section 03.015 Bottles and Jars

This subsection presents images of bottles and jars in the shape of an elephant or displaying an elephant image. There are many other examples of this category not included in this volume!

Apothecary glass jar. 13" H x 8" L. $25-40. *From the author's collection.*

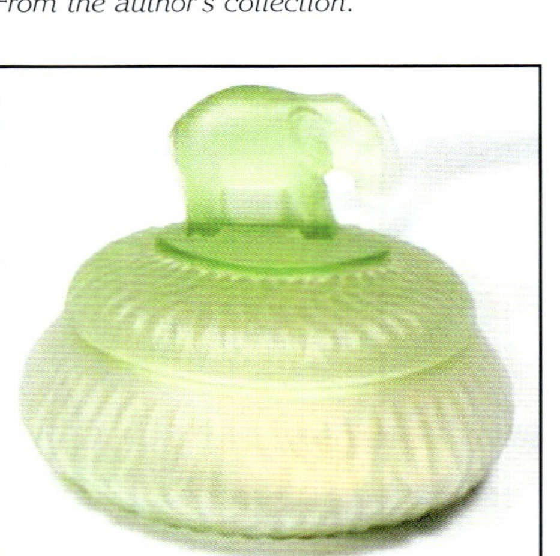

Depression glass powder jar. It has an elephant finial. 5" H x 6" L. $155. *Courtesy of Mary Lynn.*

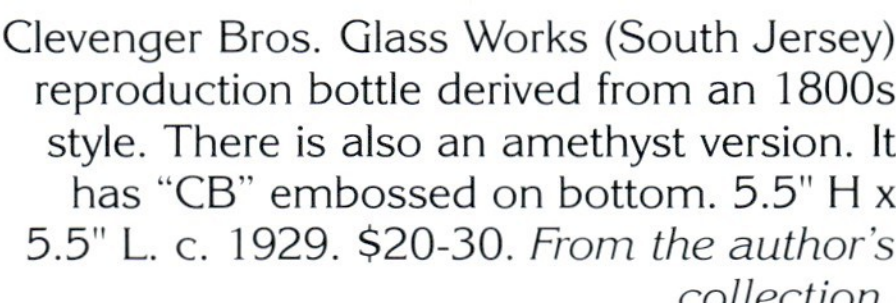

Clevenger Bros. Glass Works (South Jersey) reproduction bottle derived from an 1800s style. There is also an amethyst version. It has "CB" embossed on bottom. 5.5" H x 5.5" L. c. 1929. $20-30. *From the author's collection.*

Clear glass peanut butter jar in the shape of an elephant. It has a tin lid and has "Peanut Butter" embossed on the back of the elephant. 3.375" H x 3" L. 1930s-40s. $165-200. There are other "Jumbo" peanut butter jars including embossed bottles in different sizes (e.g., 3.5 oz., 1 lb., and 2 lbs.), made by Frank Tea and Spice Company of Cincinnati, Ohio. They are valued at $15-50.

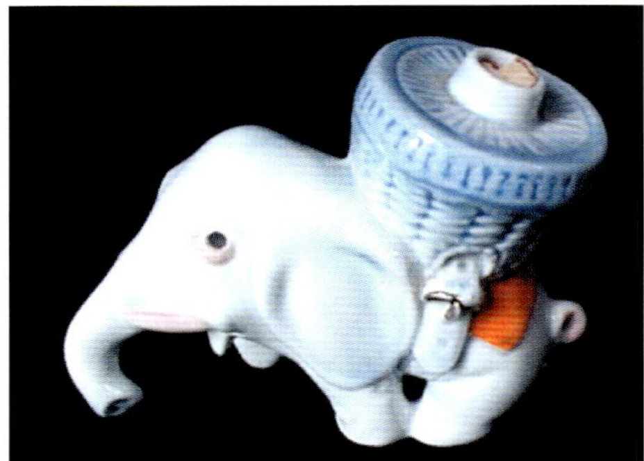

Ceramic bottle from Germany. 1920s. $40-50.

Jumbo Peanut Butter glass jar with original red and white metal lid. It is embossed "Jumbo Peanut Butter, 7 oz" on the front and "Packed by Frank Tea and Spice Co. Cincinnati Ohio." The red and white lid has a tent with an elephant graphic and Frank's Jumbo brand lettering. 3.75" H. c. 1940s-50s. $40.

Art Deco Royal Doulton flambe biscuit jar by Sung, with an elephant finial on the lid. 1930s. $2000-2500.

Unusual Schafer and Vater flask depicting an elephant standing on his hind legs spraying water over a bathing beauty. 5" H. 1920s. $750+. *Courtesy of George Speziale.*

Section 03.016 Bowls, Compotes, and Food Containers

This subsection presents images of bowls, compotes, and other containers used mostly for foods. Most of these items are made of pottery or glass, so most of the major manufacturers using those materials will be represented here, including various Staffordshire makers, majolica, depression glass makers, Dedham, and more modern makers like Lenox.

Ceramic lidded bowl marked Baronia Edelstein held in a brass filigree frame with three elephant heads with ivory tusks; it is marked Made In Germany. 5" H x 8" L. c. 1930. $75. *From the author's collection.*

Bowl incorporating a frieze of 6 elephants. Marked on the base in the mold: "Eichwald Czechoslovakia 4588 / 2." 4.25" H x 7" L. Early 20th century. $800-1000. *Courtesy of Paul Morgans and Jeff Gregory.*

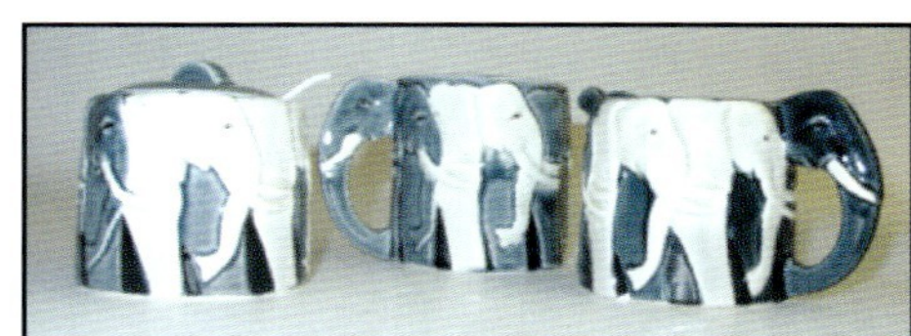

Cream and sugar containers (cup also shown), marked on the bottom "Design © Tom Taylor." 1980s. $15-20. *From the author's collection.*

This blue covered dish with rider finial is marked "Vallerysthal" raised on the inside on the bottom portion of the covered dish. These are from Lorraine, France, and made from French opaline glass. These covered dishes are extremely scarce. 5.25" H x 7.5" L. c. late 1800s - 1910. $200-500. *Courtesy of Nick Ciovica.*

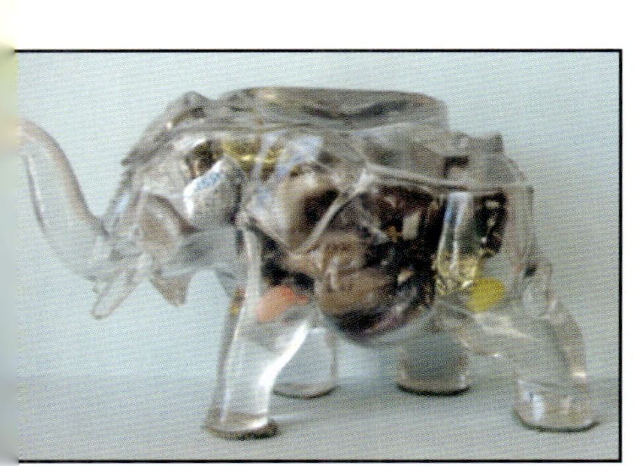

Glass covered candy dish. This dish started Adele Verkamp, who contributed many images for this book, on her elephant-collecting hobby. 4.75" H x 7.5" L. 1940-50s. $75. *Courtesy of Adele Verkamp.*

Glass compote with frosted glass elephant finial on lid. 12" H. 1880s. $750-1000.

Satin finish covered glass dish, possibly made by Athena Glass. 1920s. $1000. *Courtesy of Sharon Thoerner.*

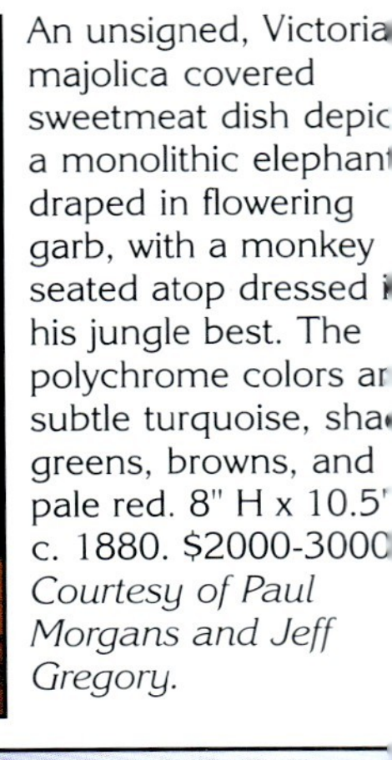

An unsigned, Victoria majolica covered sweetmeat dish depic a monolithic elephan draped in flowering garb, with a monkey seated atop dressed i his jungle best. The polychrome colors ar subtle turquoise, sha greens, browns, and pale red. 8" H x 10.5' c. 1880. $2000-3000 *Courtesy of Paul Morgans and Jeff Gregory.*

A full-lead crystal centerpiece bowl called "Elephants of the Grasslands" by Lenox. Encircling the bowl is a deeply etched design depicting an adult elephant and two young calves in typical trunk-to-tail fashion. 10" diameter. 5" H x 8" L. 1999. $125-175. *Courtesy of Adele Verkamp.*

Rare, clear covered glass dish, possibly made by Athena Glass. 1920s. $2000-3000. *Courtesy of Sharon Thoerner.*

Satin finish covered glass dish made by Co-Operative Flint Glass Co. (1879-1934). 7" H x 12" L. 1920s. $200-400. Crystal version ranges to $450 and transparent colored version ranges to $850. Tiara of Indiana made a similar dish in 1983 worth about $25 (they are no longer in business). AandA Importing has cobalt and pink versions, and Fenton produced a plain back version. This basic design was also made with ashtray, soap dish, and plain back-covers. Flower-backed covers are especially sought after by some collectors. *Courtesy of Sharon Thoerner.*

Small oval bowl from the original Fremlins Brewery of Earl St., Maidstone, Kentucky, which was taken over by Whitbread in 1967. There are imitations as the name has been reintroduced. Oval, black and red on white, showing Fremlins trademark elephant and "Fremlins Beers." Reverse has "Bristol Pottery Est. 1652" and the sailing ship trademark. 5.75" diameter. $150-200. *Courtesy of Mary Marwick.*

Flambe Royal Doulton porcelain dish is signed by Moore. It also has the marks of Sung and Noke, designers who participated in the making of this piece. Featured on each corner of the item, is an elephant head, which is Art Deco in style. The item carries the Royal Doulton Lion on Crown mark for post 1930, as it has the Made In England addition. 1.75" H x 3.25" L. post 1930. $400. *Courtesy of Antiques Centre Online.*

Flow blue ceramic soup bowl. According to Grucza, flow blue production started in the early 1800s. There are three types of flow blue: Early Victorian (1835-1850), Mid Victorian (1860-1870), and Late Victorian (1880 to 1910). It is made using cobalt oxide, which had been discovered in 1545 by Schurer. This dye sank into the porous earthenware and blurred somewhat in the glazing process. "In the 1820s it was discovered that although the blue would blur naturally, it could be made to flow by instilling lime or chloride of ammonia in the sagger while glazing." (*Jennifer Grucza, 1994* "Flow Blue China" online article) One of the classic scenes in flow blue that is found on plates and vases, includes a floral/scroll motif around edge with a middle portion that includes an Indian landscape, palace in background and the elephant with howdah in the foreground. $50-75. Early 1800s versions command $200 and up.

Section 03.017 Boxes and Caddies

This subsection presents images of all sizes and shapes of boxes made from a wide range of materials. Some other "boxes" are in the Tins category. Among the most sought-after boxes are the pottery boxes by Harmony Kingdom, manufacturers in the Limoges, France area, and Meissen, who besides making several porcelain elephant figurines, also produced porcelain boxes with elephant motifs, figures, and clasps. Many recent Limoges boxes were produced as blanks for other companies in other countries to decorate. So you may find that even though your Limoges elephant box says Limoges, France, and that it has something similar to: "Peint main" or "peint á la mein", marked on it (indicating it is hand-painted), that the box could have been painted outside of Limoges.

Paper jewelry box from Thailand. 5" H x 7" L. 1990s. $10-15. *Courtesy of Adele Verkamp*.

Wooden Burmese box; detachable rider (revealing interior) is Deva Thagyamin (God Indra) riding an Indian elephant, probably Erawan according to myth. 12" H x 9" L. c.1800s. $300. *From the author's collection*.

Porcelain box with elephant lid and brass clasp; made in Limoges area of France; these boxes range from $40-125. 2.5" H. 1990s. $100. *From the author's collection*.

Porcelain box with "S.L.A.B. England" on bottom. 2.5" H x 2" L. 1997. $35-50. *Courtesy of Adele Verkamp*.

Harmony Kingdom box called Louis, from the United Kingdom. 1.75" H x 1.75" L. 1990s. $35-50. There are many HK elephant boxes, with some valued as high as $200 or more, like Reminiscence later in this section. *Courtesy of Adele Verkamp*.

Wood box is elephant lying on stomach. 5.75" H x 10.25" L. c. 1920s. $50-75. *From the author's collection*.

Brass and pewter box. 4.5" H x 6.5" L. 1990s. $25. *From the author's collection*.

Mahogany wood box inlaid with several types of wood depicting elephants in a forest. 8" H x 16" L x 11" D. 1980s. $300. *From the author's collection*.

Three round nested brass boxes with enamel tops from India; imported by Enesco Imports. 4", 3", and 2.5" diameters, 1.75", 1.25", and 1" H. $15-20. *From the author's collection*.

Intricately detailed brass box, probably from India. 6" H x 9" L. $50-75. *From the author's collection.*

Silver box from India used to carry powdered lime, one of the three ingredients for betel nut quids. A pinch of the lime, wrapped with a betel nut in a betel leaf, was savored slowly between cheek and gum. Betel nut chewing was ubiquitous throughout Southeast Asia. 2" H x 3" L. 1870s. $100-150. Recent versions of these antique boxes are valued at $25-50. *Courtesy of www.emrimports.com.*

Depression glass dresser box with an elephant finial. 3.5" H x 3.5" W. 1930s. $65. *Courtesy of Mary Lynn.*

Oval wood box from Bali. 1990s. $15-25. *Courtesy of Maya Treasures.*

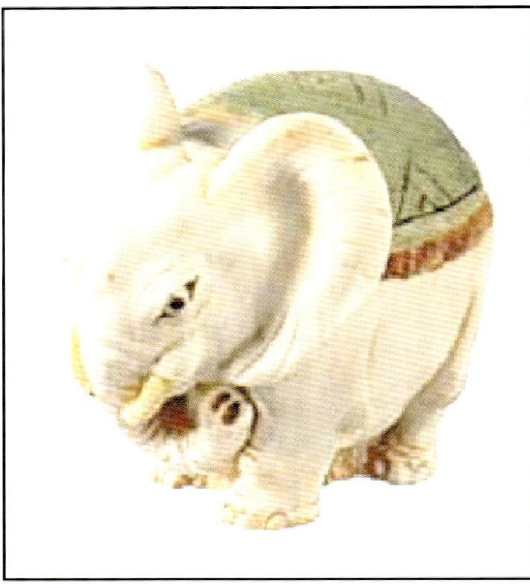

According to Harmony Kingdom's information, "Reminiscence" was sculpted and signed on the bottom by the early clay master, Peter Calvesbert for the Treasure Jest collection. Peter's signature mouse is between the hind legs. The total production run is reported as 15,957 pieces. It came in a labeled (M26 Elephant) Green HK Box and has the Old Treble-Clef Crown Stamp on the bottom of the base. 2.125" H x 2.25" L. 1993; It was retired in late 1996. $200-250; issued at a price of $35. *Courtesy of Jon Brown and www.harmonykingdom.com.*

Ivorine and Bakelite vesta box (probably used as a matchsafe). 1.5" H x 2.75" L. 1930s. $20-30. *Courtesy of Abbey Antiques.*

Unopened box of Rub-No-More cleanser made by Proctor and Gamble. 6.25" H x 3.75" L. 1930s. $75-100. There are other Rub-No-More collectibles including a watch fob, trade card, and tin sign (original and reproduction). *Courtesy of Vaughncille Griffin and Sharon Mlika.*

Section 03.018 Buckets and Pails

This subsection presents images of buckets and pails. Specifically, in this volume, I show an ice bucket and a child's pail from the well-known J. Chein Co.

Brass ice bucket from China with elephant head handles. 11" H. 1950s. $75-100. *Courtesy of Eileen.*

J. Chein sandpail and shovel. The graphics include an elephant holding a puppy on its trunk. 4.125" H x 4.25" diameter, tapering to 3.25". 1940s-50s. $40-65. *Courtesy of Nancy Schisler, Nancy's Nostalgia Antiques and Collectibles.*

Section 03.019 Buckles

This subsection presents images of belt buckles. Another well-known elephant buckle is made by the firearm maker Weatherby.

Pewter and enamel belt buckle by Bergamat Brass Works, Model #M150. 2.5" H x 3.25" L. 1990s. $15. *From the author's collection.*

Baron brass belt buckle Model #6188, from Taiwan. 2.5" H x 3.5" L. 1980. $15. *From the author's collection.*

Plastic belt buckle. 3.75" H x 4" L. 1980s. $5. *From the author's collection.*

Cartier chrome buckle on Cartier black leather belt. The buckle is in the shape of a mother elephant with her baby elephant. The word "Cartier" is stamped on the left of the buckle on the elephant's hindquarter. The buckle is of highly polished chrome that gives it a "mirror" effect. Made in the Philippines. Buckle: 1.25" H x 2.5" L. Entire belt is 41.5" L. 1990s. $140. *Courtesy of Melinda M. de Jesus.*

Section 03.020 Business Cards

This subsection presents images of business cards with an elephant image.

Adele Verkamp's business card; Adele contributed many images and knowledge to this book. 2" H x 3.5" L. 2000. N/A. *Courtesy of Adele Verkamp.*

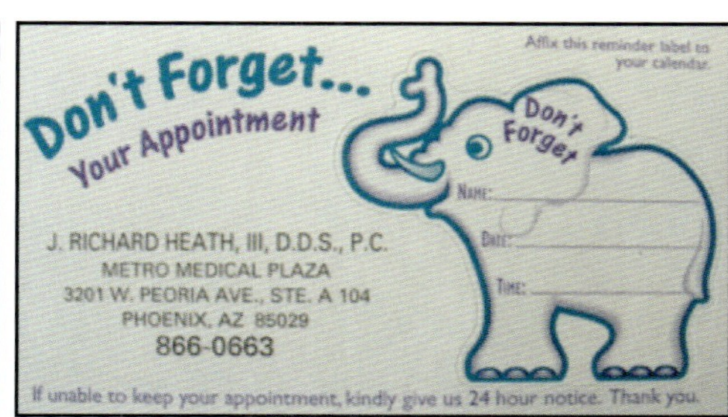

Dental appointment and business card with cute elephant graphic, used by dentist J. Richard Heath of Phoenix, Arizona. 2" H x 3.5" L. 2000. N/A. *From the author's collection.*

Section 03.021 Buttons

This subsection presents images of elephant buttons; I have seen several other buttons made from Bakelite, various metals, ivory, wood, and ceramic.

Hand-painted porcelain elephant button finished with heavy gold-trimmed details. It is an Arita button signed "TK" for Toshikane Porcelain, the early Arita backmark. This Arita pachyderm is an NBS medium, and has a self-shank. .75" H x .75" L. 1950s. $150-250. *Courtesy of Robert Selman, Blue Ridge Buttons.*

Pressed wood composition (a combination of ground up wood and other materials) button, made by the Burwood Company of Traverse City, Michigan. 1.5" diameter; larger and smaller versions were also made. 1930s. $35-50. *Courtesy of Maureen Grant.*

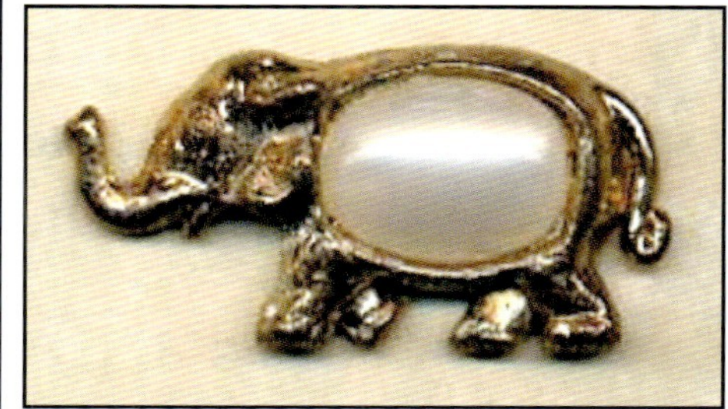

Cast metal button trimmed with moonstone "jelly belly" and elongated key shank. .5" H x 1.1" L. $20. *Courtesy of Bob at: www.blueridgebuttons.com.*

Rzanski ceramic button model #478. 1.5" H x 1.675" L. 1993. $20-25. *Courtesy of Bob at www.blueridgebuttons.com.*

Carved, incised and pigmented design on convex ivory button. Has applied knob-like shank. 0.75" diameter. $50. *Courtesy of Bob at: www.blueridgebuttons.com.*

Royal Worcester "Nelson" elephant candle snuffer, Style #05SNF304. First manufactured in the middle of the nineteenth century, this bone china snuffer is one in a new mini-series. It is hand-decorated in 22 carat gold. The elephant design has been adapted from a pattern ordered by Admiral Nelson, after he visited the Worcester Factory in 1802. 5" H. 2000. $150. *Courtesy of www.spiderweb.ltd.uk.*

Pink elephant hand painted by Edith and Allen Brooks (England) on a turquoise plastic button with a knob-like self-shank. 0.675" diameter. 1960s. $60. *Courtesy of Bob at: www.blueridgebuttons.com.*

Blue and white jasperware design (a la Wedgwood) backmarked: "SR '84," indicating design by Stella Rzanski. It has a loop-like self-shank. 1.5" H x 1.675" L. $25. *Courtesy of Bob at: www.blueridgebuttons.com.*

Bronze candle holder from Pie One Imports. 6 H x 5" L. 1990s $20. *Courtesy of Adele Verkamp.*

Section 03.022 Candle Holders and Candelabra

This subsection presents images of candleholders with elephant motifs. Metals, wood, and pottery are favorite materials with which these items are made.

Pair of silver-plate candle holders. 4.5" H x 6" L. 1970s. $50. *From the author's collection.*

Ivory-colored pair of sitting elephants with raised trunks as the candle holders. Marked on bottom with Rookwood logo and the style #6059. 4" H. 1932. $150-200. *Courtesy of Richard and Sally Burg.*

Victorian brass candelabra. c. 1900. $75.

Enesco brass candle holder, Made in Taiwan. 3.25" H x 3.5" L. 1983. $10. *From the author's collection.*

Pair of Bavarian painted ceramic candle holders in Art Deco style. 4.5" H x 4" L. 1930s. $20. *From the author's collection.*

Unusual and whimsical pottery candlesticks. The heavily scrolled base has three elephant heads with their trunks extended and intertwined at the top to hold the candle cup. The pieces are one of a kind and marked on the bottom with the Rudolstadt Volkstedt mark of L. Strauss and Sons of Thuringia. 11.25" H. c. 1900. $800-1200. *Courtesy of Pia Stratton of Pia's Antiques.*

Section 03.023 Candles

This subsection presents images of wax elephant candles. I have many of them and can never bring myself to burn them.

Gray painted candle. 5" H x 6" L. $5. *From the author's collection.*

Gray candle of sitting baby elephant holding flower in trunk. 6" H x 4.5" L. $5. *From the author's collection.*

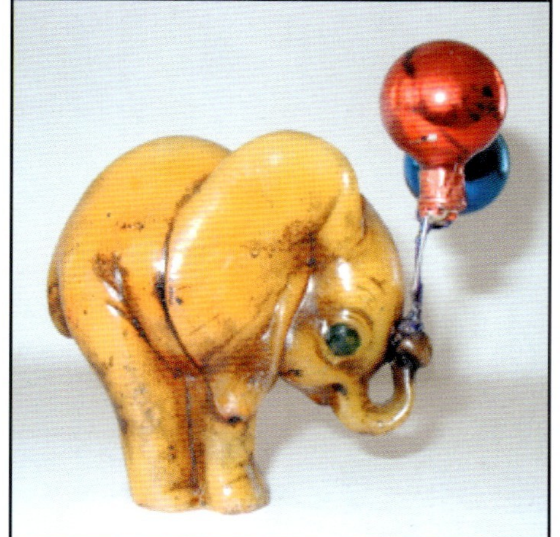

Yellow candle with foil balloons. 5.5" H x 5.5" L. $8. *From the author's collection.*

Mauve candle from Out of the Ordinary, Chicago, Illinois. 5.5" H x 8" L. $5. *From the author's collection.*

Section 03.024 Canes and Walking Sticks

This subsection presents images of canes with elephants carved into the handle. Jeffery B. Snyder says (Snyder, 1993), that the earliest canes probably date to the beginnings of humanity or beyond (perhaps some millions of years, as an aid to walking/climbing. The Egyptians, 6200 years ago, recorded the use of canes and they were symbols of power, means, faith and magic. From the 17th century through the Industrial Revolution, canes were made of many materials and although the basic design was similar, the handles were varied works of art. Ivory is a very popular handle material. Most elephants on canes are part of the handle, but some carved canes can have elephants as part of the collar or shaft as well.

Finely caparisoned German silver 2-piece cane with elephant head as handle. The cane is a one of a kind, hand made from what appears to be German silver, made in Malaysia. German silver is an alloy of various metals including silver, nickel, and copper. 5.5" L head. 36.75" overall length. c. 1900. $150-200. *From the author's collection.*

Mango-wood cane from India, with elephant head made of pewter. 36" L. 2000. $13. *Courtesy of Cost Plus World Market.*

This cane's elephant head handle is made of imitation ivory with a solid beech wood shaft and rubber toe. 37" overall length, head 5" L. 1990s. $40. *Courtesy of AJ.*

Bamboo cane, by the famous political commentator/ cartoonist Thomas Nast. He is credited with creating the GOP symbol of the elephant, and designed this cane for the 1888 Republican National Convention, and the nomination/election of Harrison. Elephant handle is silver, and the bamboo is filled, polished, and lacquered. 36" L. 1888. $225-250. *Courtesy of Patrick Weldon.*

Section 03.025 Card Holders

This subsection presents images of cardholders; these are usually made of metal, wood, or pottery.

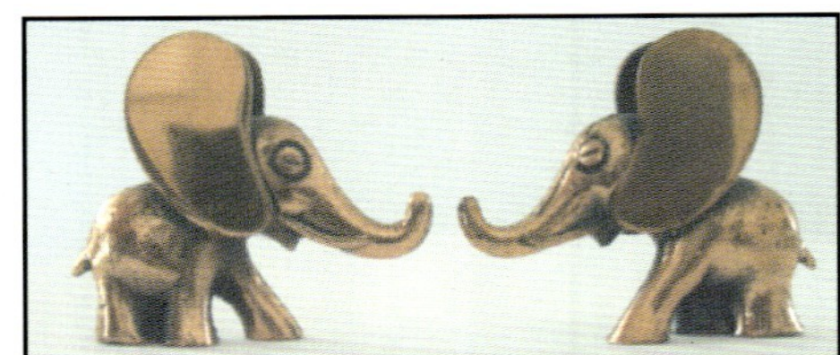

Pair of copper dinner card holders from South Africa. 1.5" H x 2" L. 1990s. $20. *From the author's collection.*

Toned pewter card holder by KandT. 1" H x 2.25" L. 1990s. $10-15. *Courtesy of Rob Faden.*

Section 03.026 CD and Album Cover Art

This subsection presents images of the albums or CDs I found that have an elephant image on the packaging. There is an "Elephant Six Recording Company" which is really a loose coalition of bands using the same label, but sadly, it does not seem to sport anything to do with an elephant.

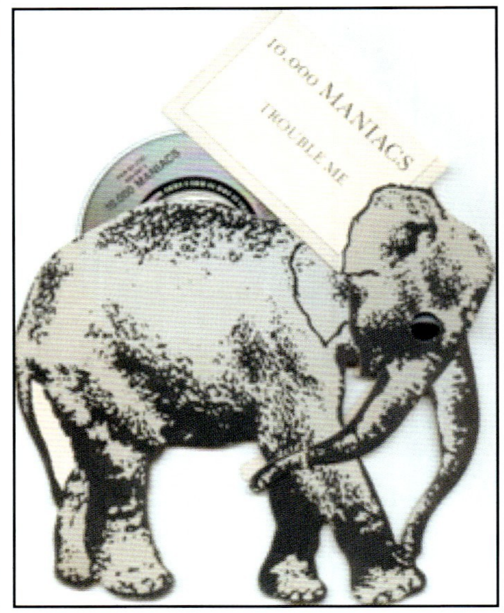

This elephant die-cut holds a CD entitled "Trouble Me" by 10,000 Maniacs, 6" x 6" for die-cut version; there is a regular 5" H x 5" L package with an elephant as well, UK release date: May 1989. $30 regular package; $100 for die-cut. *Courtesy of Ken Miller.*

Elephant artwork on a CD entitled: "The White Elephant Sessions" by a band called Jars of Clay. Original graphics by Robin Parrish; CD produced by Essential Records. 5" H x 5" L. 2000. N/A. *From the author's collection.*

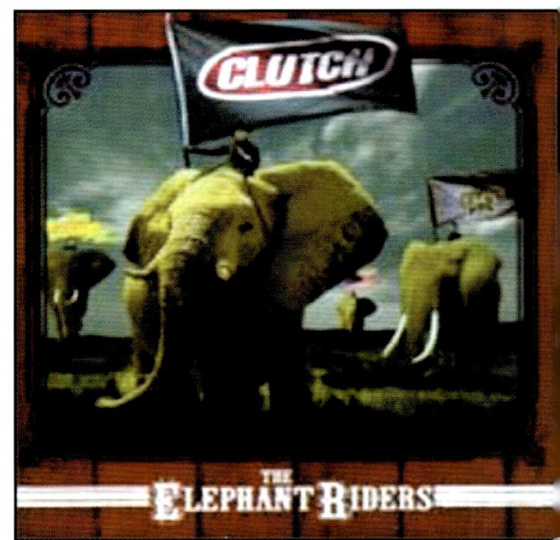

Elephant artwork on a CD entitled "The Elephant Riders" by a band called Clutch. 5" H x 5" L. 1998. $15. *From the author's collection.*

Section 03.027 Chess

This subsection presents images of elephant chess pieces. Chess pieces are usually made from ivory, various woods, or even pottery. Many chess sets sport an elephant as one particular type of chess piece, usually the rook, but a knight riding an elephant is also common.

Ivory chess pieces; rooks from east Asian region. 3" H x 1.5" L. 1950s. $15-20 each. *From the author's collection.*

Bone chess set pieces, all elephants royally caparisoned. 2-4" H. 1990s. $150. *Courtesy of beekay@bee-kay.com.*

Burmese wood elephant chess piece; knight. 3.5" H. $30. *Courtesy of Brigitte and George Whiteman.*

Section 03.028 Cigar Cutters and Dispensers

This subsection presents images of cigar cutters and dispensers. According to Terranova and Congdon-Martin, 1996, the earliest cutters were made by Erie Specialty in the 1880s, and Brunhoff Manufacturing Co. in the 1890s. There are certainly elephant cutters going back to 1900 and perhaps earlier.

Bronze and ivory cigar cutter. 1.5" H x 7" L. c. 1920s. $400. *Courtesy of Dan Morphy, Adamstown Antique Gallery.*

Antique cigar cutter depicting an elephant made in pewter with the characteristic round belt engraved, and the brass cutter on top,. The cutter is pushed down to cut the cigar, then lifted to expel the tips. Two sized holes. See *Antique Cigar Cutters and Lighters, pp.36-37* for similar examples. 4" H x 6" L. $400-500. *Courtesy of Buster Berntson.*

Bronzed cigar cutter. 2.25" H x 5.25" L. $75-125. *From the author's collection.*

Rare bronze cigar cutter. $800-1200.

Unique cigar box and dispenser. When you pull down on the back lever, the box opens and the elephant picks up a cigar with his trunk. The elephant is made of leather over metal or wood and the box is covered with leather. It has a sticker on the bottom reading: "Made in Czechoslovakia" and on the side it says" Great Lakes Expo 1936." $200-300. *Courtesy of Lois Wright.*

Section 03.029 Cigarettes and Cigarette Holders

This subsection presents images of cigarette packages and holders/dispensers. In Righini and Papazonni, 1998, the authors say that tobacco has been used since the 15th century and cigarettes since the mid-1700s. In the mid-1800s cigarettes were mass-produced and packs of cigarettes became popular around World War I, so it is no surprise that elephants and cigarettes came together in Bear's "Elephant Cigarettes" in the 20s and 30s, made by Thomas Bears and Sons, UK. Two Jumbo brand packs are also shown. Another elephant image is found on Bangle cigarettes made by U.L Allen Ginter, 1965. In 1950 Honeydew cigarettes sported an elephant.

Jumbos cigarette pack, by the Moon Light Tobacco Co., a subsidiary of RJR. 3.25" H x 2.5" L. c. 1995. $10. *From the author's collection.*

Cast iron cigarette dispenser. The tail is the crank handle, when turned it dispenses the cigarette from the elephant's body. 5" H x 7.5" L. 1920s to 1930s. $300-350. A 1930s version with an ashtray as the base is worth about $400-600. *Courtesy of Jeff and Gina Fuhrmann.*

Hand-carved wooden cigarette box with intricate interior to hold individual cigarettes; probably from India. 7" H x 7" L. 1985. $20-30.

Bangle cigarette pack by Allen and Ginter of Richmond, Virginia. 3.25" H x 2.5" L. 1990s. $20-45. *Courtesy of J. Shattelroe.*

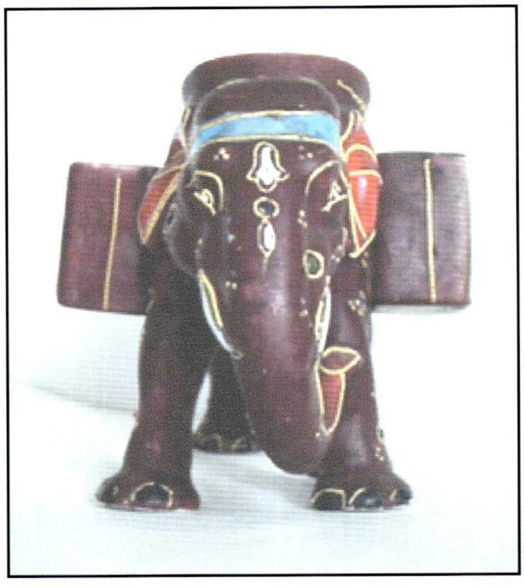

This elephant is unmarked but is probably made by Morton Pottery of Illinois (1877-1971). The elephant has a rectangular box on each side intended to hold packages of cigarettes. The clay is tan and has a grayish/purple drip glaze over a semi-transparent tan/flesh tone. 4" H x 6.25" L. c. 1950s. $30. *Courtesy of Janet Wojciechowski.*

Jumbo brand cigarette pack made by the Larus and Brother Company, subsidiary, Reed Tobacco. Larus and Brother was bought by Rothmans of Pall Mall Canada Limited in 1968. The continued production of their tobacco and cigarette brands, including Jumbo, was contracted out to the Liggett and Myers Tobacco Company. 3.25" H x 2.5" L. c. 1920s. $150-200. *Courtesy of James A. Shaw.*

Section 03.030 Circus Posters and Programs

This subsection presents images of circus posters and programs. Most circus posters have had elephants on them. Posters can be made of heavy paper or even heavier cardboard or card stock, especially earlier ones designed to be stapled to telephone poles and other public venues. Original, older posters can range to the thousands of dollars (like the Carter Vanishing Elephant act poster), so many collectors choose to buy a reproduction which is true to the original image and delights nonetheless.

Ringling Brothers and Barnum Bailey circus program. 11" H x 8.5" L. 1955. $15. *From the author's collection.*

A. G. Kelly Miller Bros. has been putting on circuses since 1938. This circus poster is on heavy card stock. 14" H x 11" L. c. 1950s. $10. *From the author's collection.*

Ringling Brothers and Barnum Bailey's Greatest Show on Earth poster, Reproduction, StruBridge Lithographers. 10" H x 14" L. 1990s. $7. *From the author's collection.*

Section 03.031 Clocks

This subsection presents images of elephant clocks. Some of these date back 200 years or more, especially bronze/gilded French clocks, which are known for being very ornate (and quite valuable). One of the most sought-after elephant clocks is the Swinger by Junghans. A similar swinger was made in Japan by Linden in the 1960s.

Leonardo Collection clock from China; elephant is composition. 9.5" H x 11.5" L. 1990s. $50-75. *Courtesy of Ted Capell.*

White Lenox China Treasures clock model #2583. 5" H x 9" L. 1990s. $125-150. *Courtesy of Adele Verkamp.*

Ceramic clock by Capodimonte of Italy, signed by artist. This elephant not only doubles as a quartz crystal-run clock, but the top can hold a bowl or flower arrangement. 14" H x 16" L. 1990s. $150-200.

Bronze console clock with an elephant supporting mechanism and other bronze figures. Clock base has inset painted ceramic image. 20.5" H x 14" L. 1890s-1920s. $250-300. *From the author's collection.*

Celluloid Ivorine elephant clock. Impressed "Tuskeloid" within diamond shaped border on base; Made in U.S.A. 5" H. c. 1925. $60-75. *Courtesy of Bob Randolph.*

Bronze clock with wind-up movement by Gustav Becker who founded his clock factory in 1848 in Silesia and continued through the 1890s. He won a design award in 1852 and a Medal of Honor, which became part of the trademark and is stamped on the back of the movement. 13" H x 9" L. late 1800s. $1200-1700. *Courtesy of Archivator (Budnitski Vladimir).*

Rare 19th century signed, gilt and bronze, elephant mantel French clock. Eight-day movement with striking of a bell on the hour and half hour. Outside counterwheel. Movement is stamped "Japy Freres and Cie," "Med. D'Honneur." White enamel dial with blue Roman numerals. Seated oriental man holding a parasol above the clock dial being carried on an elephant's back. Mounted on a scroll base. 16.5" H x 14" L x 5" D. 1800s. $2500. *Courtesy of Stella.*

French bronze and gold elephant clock named "La Horloge E'le'phant Elegant D'or" (The Golden Elephant Clock). Elephant is standing on a golden base of foliage, supporting the clock with its porcelain dial. The trainer is a bronze Asian boy holding an elephant hook. The movement is by Marti and Co and runs for 8 days and strikes on the hour and half hour. 15.5" H x 13" L x 7.5" D. 1890s. $4000-5000. *Courtesy of Ted Wittmann.*

Art Deco style, nickel, 3-piece clock set carried by 2 elephants mounted on a beige marble base. Probably made in France or Germany. The movement is German (Junghans) but this does not necessarily mean the set was made in Germany, as Junghans movements were used all over Europe. 6" H x 8" L. 1890 to early 1900s. $350-500. *Courtesy of Gijs van der Plas.*

Junghans German bronze "swinger" clock, so named because its unusual movement is the whole clock mechanism swinging from the elephant's trunk. Also referred to as the Junghans Mystery Clock. Genuine Junghans swinger clocks have a movement made by Junghans and also put "Junghans" on the clock's face; reproductions, which could use the Junghans movement, do not put that name on the face. 12" H x 12" L. 1890-1900. $600-1000 for originals, $150-300 for reproductions. *Courtesy of Lauren Wentz.*

Section 03.032 Clothes

This subsection presents images of wearable elephants (jewelry is also wearable and has its own section). Probably the most common article of elephant clothing is the T-shirt. There are probably thousands of different elephant T-shirts; just think of all the zoos in the world that have one or more T-shirts as souvenirs!

Cap with embroidered needlework, sequins and appliqué from Kalaga region of Thailand. c. 1970s. $20. *From the author's collection.*

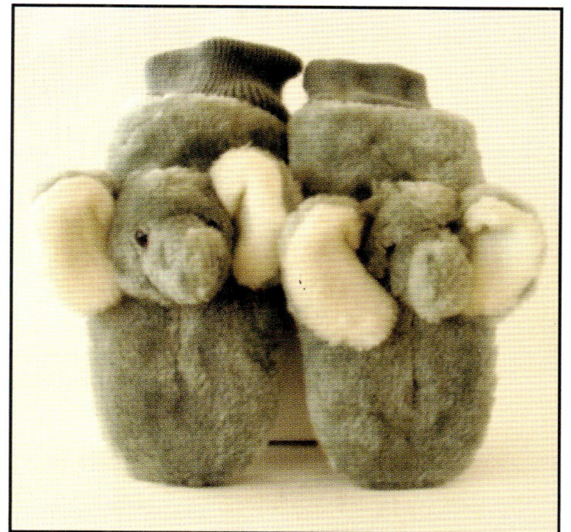

Gray children's mittens. 1980s. $10. *From the author's collection.*

Tan leather women's shoes with brass elephants on heels by Classique in China. Heel: 1.5" H X 2.5" L. 1990s. $50. *Courtesy of Adele Verkamp.*

Gray children's slippers by MIT. 1980s. $10. *From the author's collection.*

T-shirt, black with hand-painted gold colored elephants by India Ink Paintworks. 1990s. $20. *Courtesy of Adele Verkamp.*

Silk tie by Rene Chagall. 1999. $10-20. *From the author's collection.*

T-shirt from Elephant Bar, Bombay, India. Large. 1999. $10-20. *Courtesy of Stephanie Glili.*

Section 03.033 Coasters

This subsection presents images of coasters with images of elephants. Many beer makers with an elephant logo or advertising brand made items sporting their elephant image, including coasters. Coasters are usually made of heavy paper or cardboard, but other materials include wood, pottery, marble, etc.

Delirium Tremens coaster, voted best beer in the world. 3.5" H x 3.5" L. 1990s. $5. *From the author's collection.*

Various Fremlins beer coasters by Fremlins Brewery, before the brewery was bought out by the Whitbread Brewing Company in 1967. 3.5 H x 3.5 L. various dates pre-1967. $5-20 each. *Courtesy of Mary Marwick.*

Large coaster for Imported Elephant beer. 6" D. 1990s. $6. *From the author's collection.*

Leather coaster from MMSSL in South Pittsburgh, Tennessee. 3.5" D. 1990s. $7-10. *Courtesy of MMSSL.*

Ceramic coaster with a fired gold rim by New York-based manufacturer, Marians Gifts and Ceramics. 4.2" D. 2000. $20. *Courtesy of Antonio.*

Section 03.034 Coins, Tokens and Medals

This subsection presents images of world coins, tokens, medals and currency. There are hundreds of coins with elephant images, and I've presented a few spanning the ages from ancient coinage (i.e., from approximately 1500 A.D. to 2500 B.C.) to current coins. Many of the older elephant coins used an elephant image to indicate conquest or power, as in the case of the Caesar Denarius, where the elephant symbolizes Caesar crushing his enemies (a snake beneath the elephant's feet).

Other issuing countries used elephants because they were native to their lands. There are also several currency notes that have elephants on them, and certain commemorative medals have featured elephants. Because numismatics is another long-held collecting interest of mine, I intend to place many more elephant coins into my next volume! By the way, this is an excellent way to specialize your elephant collecting - choose a category that also represents another of your "obsessions." I have put together an extensive list of elephant coins that I can send to you; contact me via email.

Ten dollar gaming token made of .999 silver from the Nugget Casino, in proof condition. 1.75" D. 1996. $15-20. *From the author's collection.*

Republic of Liberia two-cent proof coin from 8-coin proof set (of which 4 have elephants on them) made by Franklin Mint. Krause #12a. 1.25" D. 1978. $2-5. *From the author's collection.*

Great Britain copper half penny (farthing) in extra fine (EF-45) condition. Lady Godiva on horse on obverse. Also known as Tradesman, made by many private entities because of a shortage of small coinage in the late 1700s. 1792. $10. *From the author's collection.*

One ounce .999 fine silver ingot from the U.S. Silver Corp, 1974. $10. *From the author's collection.*

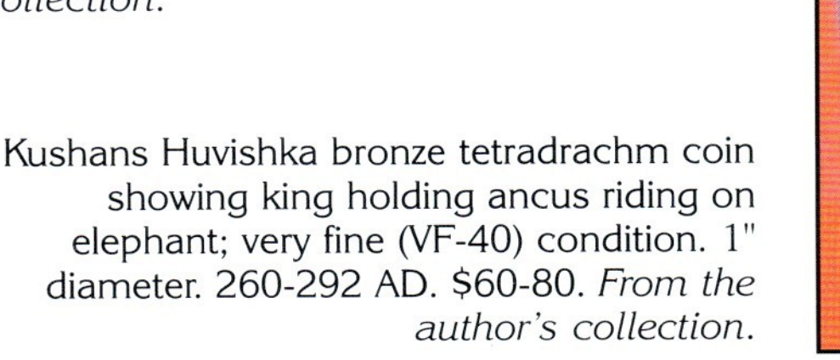

Kushans Huvishka bronze tetradrachm coin showing king holding ancus riding on elephant; very fine (VF-40) condition. 1" diameter. 260-292 AD. $60-80. *From the author's collection.*

Bronze or brass Hoover Lucky Pocket Piece token in very fine (VF-40) condition. 1" D 1932. $10 ($25 in MS-60). *From the author's collection.*

Somalia 10 centesmi copper coin in Mint State (MS-63) condition. Krause #3. 1.375" D. 1950. $5. *From the author's collection.*

Elephant half penny (farthing) London Elephant Token, thick planchet, certified by PCGS as MS-63 brown. This coin is quite rare in this condition, as only 1 coin is certified in MS64, and 8 coins are certified in MS63. 1" D. c. 1666. $3500. *From the author's collection.*

Julius Caesar silver denarius depicts an elephant trampling a serpent, said to symbolize Caesar defeating his enemies, in extra fine (EF) condition. 49-48 B.C. $1000-1500. *Courtesy of Classical Numismatic Group's June, 1999 Auction Catalog. (E2D4).*

Two franc brass coin from Belgian Congo, Krause #25, in MS-60 condition. 1943. $35. *From the author's collection.*

Ceylon bronze-over-copper 1/48th Rupee "pattern" coin (normal runs were copper) in certified Proof-63 condition. Krause #KM-75. 1802. $150-225. *Courtesy of Anthony's Stamps and Coins.* (E2D4).

Silver-colored aluminum doubloon from the 1995 Mardi Gras in New Orleans; issued by the Zulu Special Aid and Pleasures Club. 1.5" D. 1995. $20. *From the author's collection.*

Sterling silver token minted for the 20th NENA (New England Numismatic Assoc.) conference in Boston, Massachusetts. 1964. $40-50 in MS-63 condition. *Courtesy of Duane.*

Brass token made for Christmas by Harry Herzberg for his Circus Fans Association. 1.3" D. 1931. $100. *Courtesy of James E. Kattner Collection.*

French colonial Laotian medal depicting the three-elephant head design of the Order of a Million Elephants and White Parasol. According to Ed Emering author of "Orders, Decorations and Medals Webring": "...the Order was created by Sisavang-Vong, the King of Louang-Prabang on May 1, 1909 to recognize exceptional military and civil service. It is awarded in five classes (Grand Officer, Grand Cross, Commander, Officer, and Knight); this one is a Knight class medal; other classes have different designs. The uniface insignia is composed of three white enamel elephants with red enamel headpieces suspended beneath a golden crown, surmounted by a stylized parasol. The Knight's badge is 56mm and the other levels, 62mm. It was widely awarded to French officials and military personnel, including those who participated in the Franco-Laotian Resistance against the Japanese in 1944 and to members of the French Military Mission and the French Expeditionary Corps, which participated in the Laotian War." Size of medal only: 2.25" H. 1950. $175. *Courtesy of Ycheche.*

Section 03.035 Computer Related

This subsection presents images of elephant items related to computers. There are several mouse pads I know of, many advertisements and computer magazine covers (see those categories for examples), and other items. There are many animated or static elephant images I have seen on web sites.

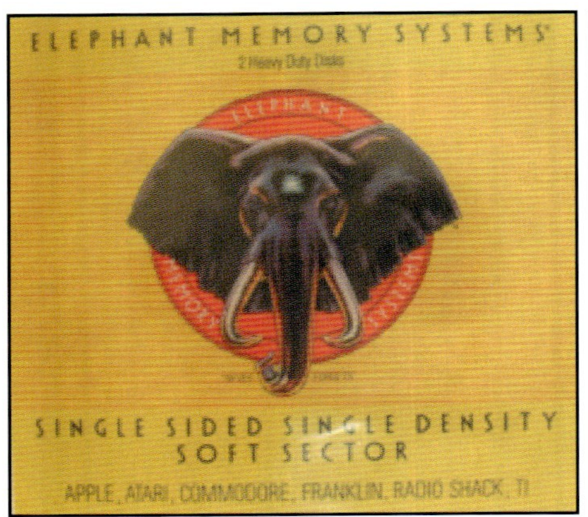

Elephant Memory Systems floppy disk pack holds 2 5.5" disks and was made by Dennison Mfg. Co. 5.75" H x 5.75" L. early 1980s. $5. *From the author's collection.*

Mouse pad by Platinum Education. 8" H x 8" L. 1990s. $3. *From the author's collection.*

Twin Winton's Sailor Elephant ceramic cookie jar, marked "Twin Winton." They also made a matching pair of salt and pepper shakers valued at about $30. 11.5" H x 9" L. c. 1960. $40-50. *From the author's collection.*

IGA ceramic cookie jar; a first edition, as determined by the blue hat; the green hat model is the second edition. 12" H x 14" L. 1998. $25-35. *Courtesy of Claude Reed.*

Section 03.036 Cookie Cutters

This subsection presents images of elephant cookie cutters. Probably the most sought-after cutters are the vintage (c. 1940s-50s) green or red wooden-handled variety.

Large tin cookie cutter with "Handcraft From Europe West Germany" imprinted on inside. 7" H x 10" L. probably 1960s. $35. *From the author's collection.*

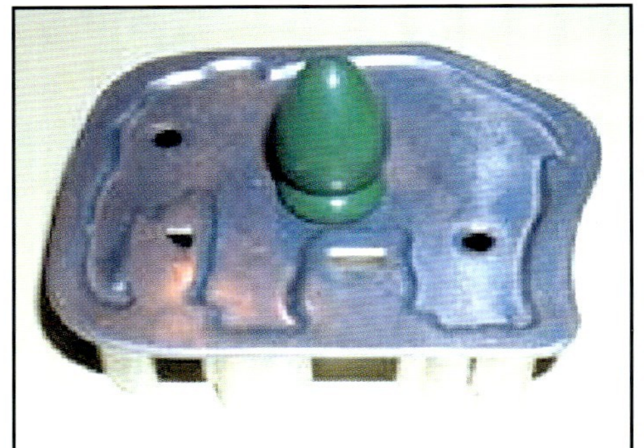

"Green Knob" cookie cutter, one of a set of three with different color knobs. 2.875" H x 3.375" L. 1940s-50s. $30-40. *Courtesy of Kim Miller.*

Section 03.037 Cookie Jars

This subsection presents images of elephant cookie jars made by well known potteries like Twin Winton and Shawnee or a contemporary store brand like IGA. The Disney section also has cookie jars.

Relatively rare elephant sailor ceramic cookie jar from American Bisque, who made such jars from 1930-1973. 1950s. $50-75. *Courtesy of Up Your Attic Antiques Of Canada.*

Ceramic cookie jar from China. 1970s. $25.

Original ceramic "Lucky" cookie jar with gold trim and green collar band (red band is rare) by Shawnee; these were reproduced in the 1970s and are valued at about $30. 11.75" H. c. 1940s. $200-750 depending on decoration; green collar $350-500. *Courtesy of John C. Horn.*

Ceramic celadon cookie jar by Ludowici. This mold was copied by APCO and ABCO. 12" H. $75-150

Ceramic cookie jar by Metlox. Designed by Bob Allen and Mel Shaw, this jar was in the first group made by Metlox from the 1950s through 1963. $750 if number 33 on the bottom, $200-300 otherwise.

Section 03.038 Cups, Glasses, Steins and Mugs

This subsection presents images of drinking vessels. There are a great many elephant cups and mugs out there, and I know of a few more steins I will try to get into the next volume. In addition, see the Shot Glasses and the Pink Elephant sections.

Elephant Red 1-pint beer glass. 6" H x 3.5" L. 1990s. $5-8. *From the author's collection.*

Budweiser Endangered Species African Elephant stein, by Anheuser-Busch Inc., A.B. St. Louis, Missouri. Made in Brazil by Ceramarte. 6.5" H x 5.5" L. 1991. $40-50. *From the author's collection.*

"Pounder" glass from US Navy's USS Camden AOE-2, read "The Powerful Pachyderm," the Camden's mascot. 6" H x 3.5" L. 1960s. $10-12. *From the author's collection.*

Rossini ceramic cup from Japan. 3" H x 4.5" L. 1980s. $5. *From the author's collection.*

Ceramic, hand-painted saki cup, from a set of eight. 3.25" H x 4.5" W. 1940s-50s. $5-7. *From the author's collection.*

Vienna cup and saucer, hand-painted with a gray camaieu elephant. This cup belonged to the famous Imperial Vienna set of "Vertues." The "vertue" of elephant, inscribed in French, is: "la Magnanimit." The saucer is hand-illustrated with an elf playing a flute in a field of flowers. Cup 2.3" H x 2.3" diameter. Saucer 5.25" diameter. c. 1800. $150-175. *Courtesy of Danielle Cordier.*

Section 03.039 Desk and Office Accessories

This subsection presents images of certain desk accessories including: pencil and penholders, letter openers and calendar holders. Other desk-related items have their own categories: Pencils, Pens and Sharpeners, Tape Dispensers, and Inkwells.

Brass pen holder with walnut base and marble elephant. 3.5" H x 7" L. 1970s. $10. *From the author's collection.*

Copper-colored cast calendar holder. Patricia Robak has a very similar white metal calendar with a figural dog; she says that it came as a souvenir from Luray Caverns, Virginia. See *Robak, 1999*. 3.5" H x 6" L. 1930-1950. $30-45. *From the author's collection.*

Vintage art deco amethyst glass blotter with original pad and retaining clips. Made by Imperial, this is a part of a desk set that includes a large paperweight with elephant handle. 3.5" H x 7" L. 1940-50s. $30-40. *Courtesy of Suzanne Harris of Valleytreasures.*

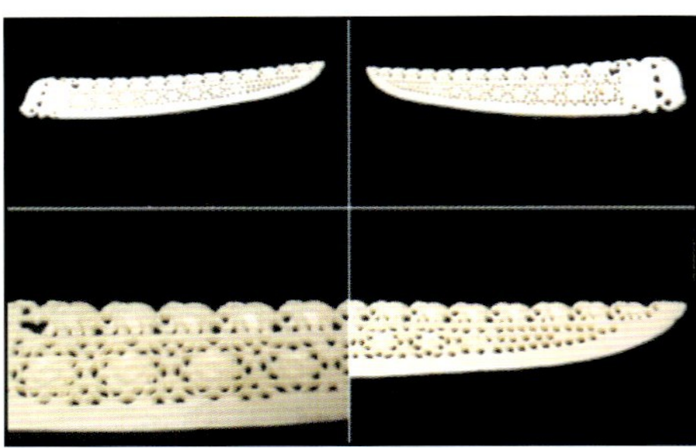

Ivory letter opener with cascade of elephants running the length. 7" H x 1.25" L. $50. *Courtesy of Henry J. Hebing.*

Section 03.040 Die-cuts

This subsection presents images of die-cuts. These are usually cardboard, heavy paper, or tin images that are cut out according to the outline of the image. These were used mostly for advertising, from the late Victorian era to the present. Many die-cuts, like the sought-after Oyster Die-cut, (where several elephants are playing cards around a table), are in great demand. Some tin die-cuts are in the Signs category.

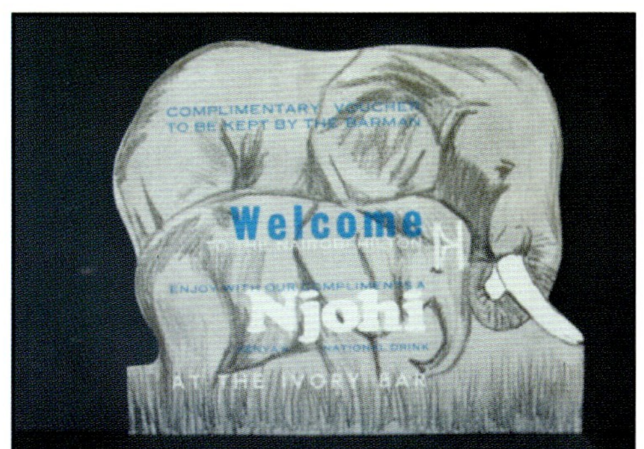

Nairobi Hilton Ivory Bar drink voucher die-cut. 5.25" H x 4.5" L. c.1960s. $5. *From the author's collection.*

Elephant Victorian die-cut with "B" and "652." 5" H x 7.25" L. 1880s-1910. $15-25. *From the author's collection.*

Advertising Elephant Milk Biscuits round die-cut; the elephant was a trademark of Holmes and Coutts, a New York company that was one of the predecessors to the National Biscuit Company, AKA Nabisco, prior to 1902. $40-50. *Courtesy of Kathy.*

Die-cut advertising peanuts made by the Superior Peanut Company of Cleveland Ohio. 2" H x 2" L. $30-50. *Courtesy of Ron Koenig.*

Die-cut of Toung Taloung, which means Gem of the Sky (the name of a tea). Toung was P. T. Barnum's white elephant he purchased in 1884. 6" H x 8" L. c.1890s. $20-35. *Courtesy of William A. Smith.*

Large die-cut advertising for the Great Atlantic and Pacific Tea Company. Has grandmother on a trip in India riding on the back of a white elephant. She says "The Great Atlantic and Pacific Tea Co.'s celebrated Teas and Coffees have been my solace through life." Signed Grandmother. At the bottom it says "Toung Taloung," the name of the tea. 10.5" H x 8" L. 1894. $15-20. *From the author's collection.*

Die-cut showing the death of Jumbo, the largest elephant, at the time touring with P.T. Barnum and Bailey Circus. On his second visit to St. Thomas, on September 15, 1885, Jumbo had a fatal collision with a Grand Trunk locomotive. 4.25" H x 6" L. c. 1885. $15. *From the author's collection.*

Section 03.041 Disney

This subsection presents images of Disney's elephant characters, the most famous of which is Dumbo the Flying Elephant. Disney created several elephants besides Dumbo (1941), including for example: Elmer (1936), Mickey's pet elephant Bobo (1936), Hathi the baby elephant in *Jungle Book*, Tantor from *Tarzan*, and Stewie from *Marsupilami*. Note that I do not count Eleroo, because it is a fantasy combination of an elephant and a kangaroo. Of course, this category could expand into a whole volume by itself, as there are countless subcategories of things Disney/Dumbo. These characters were licensed over the last half-century to various companies seeking the cachet of association with one of the most successful brands ever, so there are hundreds of Disney elephant collectibles!

One recent example is the creation of "Vienna Bronzes" of certain Disney Fantasia 2000 characters by Salzburg Creations. Animation cels are one type of Disney collectible I will include in later volumes; they range from the hundreds to thousands of dollars. Disney also produced elephant documentaries such as the one-hour "Elephant Journey," (2000) by filmmaker Adrian Warren, made as part of the nature series "New True Life Adventures." The film shows the elephants' migration through northwestern Namibia.

Disney porcelain figure. $5.

This dancing ballerina elephant figurine is one of a group of three produced by Vernon Kilns under Disney license for the movie Fantasia and is also pictured in *Frick and Hodge*, p. 46. The number "26" is impressed on her base along with the ink stamp: "Disney Copyright 1940 Vernon Kilns U.S.A." 5.5" H. 1940s. $300-600. *Courtesy of PostScript Vintage Collectibles.*

Cookie jar in likeness of Dumbo, with Timothy Mouse on his cap; made by Treasure Craft. 11.25" H x 9.5" L. $25-35.

Elmer, the Disney elephant figurine in bisque porcelain. $275.

Composition Disney Dumbo figurine. On his right leg is marked: "Cameo Doll Prod." He has felt-like ears and button eyes. His head is held to the body with a piece of rubber connected to wire hooks and it moves to any position. His trunk is connected the same way and is also positionable. 1941. $150-300. *Courtesy of Nino.*

Disney Dumbo with original label "C. Disney, Dumbo, American Pottery Co. Los Angeles." On the bottom is incised "N41." 5.25" H x 6" L. $75-125. *Courtesy of Betsy Kirk.*

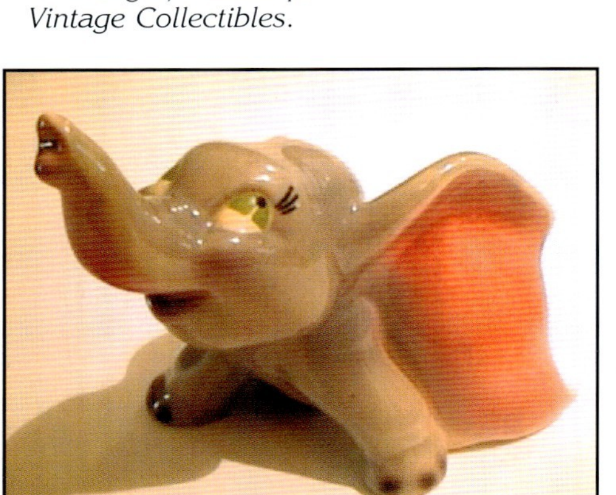

Ceramic Dumbo figurine from Modern Ceramic Products (MCP), one of the first companies in Australia licensed to make pottery figurines of some of Walt Disney's famous characters in the mid 1950s. 4.25" H. 1950s. $90-100. $120-150 with original foil. *Courtesy of sueclay@iweb.net.au.*

Goebel's Disney Dumbo with original sticker that reads: "Walt Disney, Dumbo, Character WDP.Ffm. C." On the green base is incised "Disney" and on the bottom is "DIS122." Also on the bottom are stamped, in ink: "Germany," a full bee and "R" and "C," and another Goebel mark. 4" H x 4.5" L. 1950-55 (based on Full Bee mark). $125-180. *Courtesy of Betsy Kirk.*

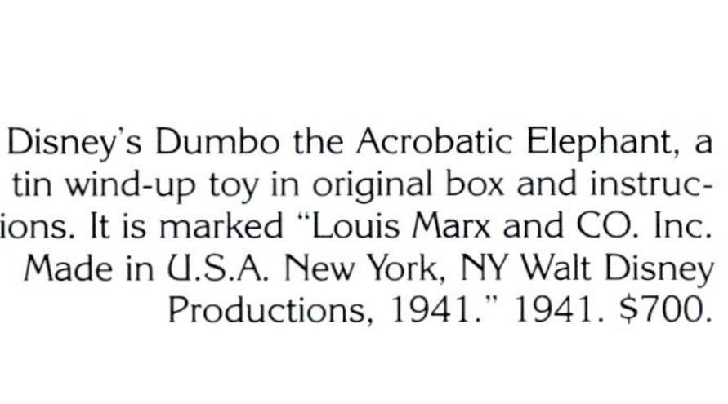

Disney's Dumbo the Acrobatic Elephant, a tin wind-up toy in original box and instructions. It is marked "Louis Marx and CO. Inc. Made in U.S.A. New York, NY Walt Disney Productions, 1941." 1941. $700.

Section 03.042 Door Hardware

This subsection presents images of different door hardware, including: door stoppers, handles, and knockers.

D.K. Miller Lock Co. metal lock with an elephant impressed on surface. There are 3 different sizes of these Miller locks. 2.5" H x 1.5" L. 1930-32. $20. Yale and Towne bought Miller out in 1932, but made the same style lock until 1950. *From the author's collection.*

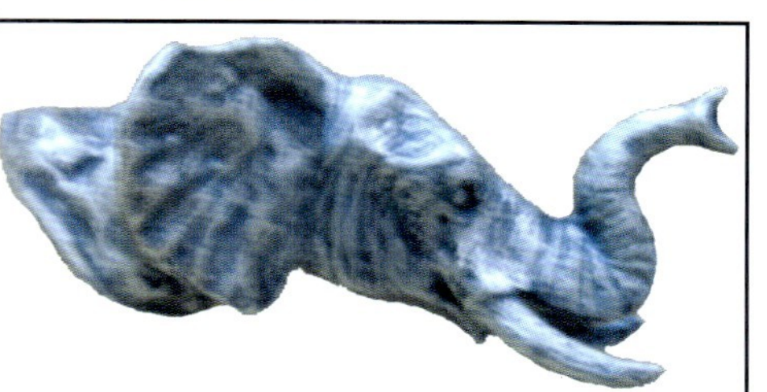

Pewter Door Knob, Model #147-1. 2.5" H x 5" L. 2000. $38. *Courtesy of The Great Indoors.*

Bradley and Hubbard cast iron doorstop; "BandH 7755" stamped on the back. Golden original paint and patina. Doyles in New York recently offered one in the $500-700 range 1.5" H. c.1910. $400-800.

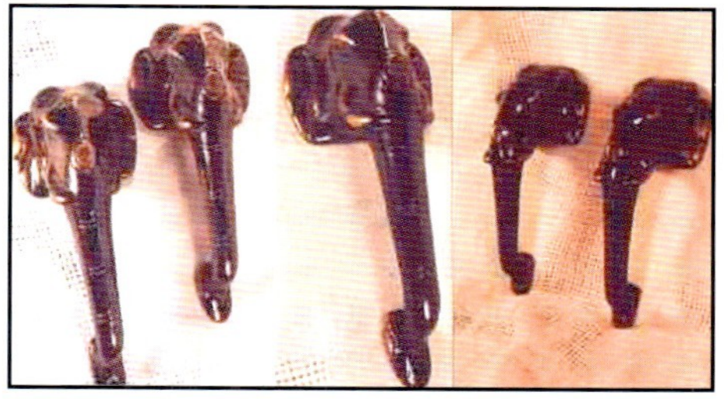

Pair of large bronze door handles. They are manufactured via the lost wax method in Nepal. 1980s. $60-100. *Courtesy of Diane Mandle.*

Cast iron elephant and palm tree doorstop. 13.5" H x 10" L. c. 1920s. $75-100.

Victorian cast iron door knocker. c. 1900. $75-125.

Hand-made Bronze door pull. 8" H. c. 1900. $40-50. *Courtesy of Tim Murphy.*

Section 03.043 Egg Cups

This subsection presents images of eggcups.

Porcelain egg cup. 2.75" H x 2.5" L. 1940s-1950s. $8. *Courtesy of Adele Verkamp.*

Egg cup made by the Keele Street Pottery in England. All paint is under the glaze. 3" H x 2.5" L. 1940s. $25-45. *Courtesy of Leona Gonzales.*

Section 03.044 Ephemera

This subsection presents images of paper collectibles with elephants, not included in other categories like stamps, labels, or postcards. This includes menus, stock certificates, letterheads, etc.

Busch Gardens cardboard brochure holder. 17" H x 9" L. 1980s. $8. *From the author's collection.*

Sunland RV Resorts brochure. 8.5" H x 4" L. 1999. $5. *From the author's collection.*

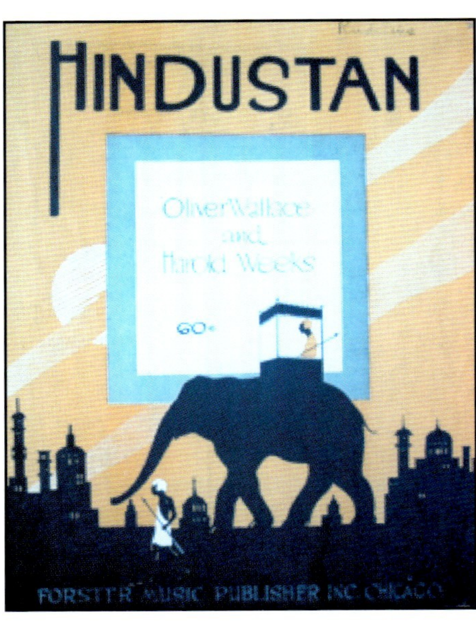

"Hindustan" Music Booklet CVR Forster Music Publisher, Oliver Wallace and Harold Weeks. 13.5" H x 10.5" L. 1918. $10. *From the author's collection.*

Menu from the Royal InterOcean Lines ship restaurant. March 10, 1954. $15-25. *Courtesy of Katie Achaval.*

Stationary pack by Freelance, #1420. 1978. $8. *From the author's collection.*

Origami: U.S. Dollar Bill folded into an Elephant. 1.5" H x 3" L. 2000. $1. *From the author's collection.*

Missouri map by Conoco. 4" H x 9" L folded. 1940s. $5-8. *From the author's collection.*

Elephant calendar published by Brown and Root. Photographs by F. Lanting. 12" H x 12" L. 1999. $10. *From the author's collection.*

Section 03.045 Fans

This subsection presents images of hand-held fans with elephants on them.

Pewter hand fan with Indian elephant head as handle; made by Reed and Barton. 6.75" H x 9.75" L. c. 1940s. $50-60. *From the author's collection.*

Wood die-cut political fan graphics by Lawson Wood. Reverse: Peoples Supply Co. Georgetown Ohio printed by Brown Bigelow. 8.5" H x 8" L. 1936. $15. *From the author's collection.*

K and M International paper foldout fan. 9" H x 10" L. 1970s. $5. *From the author's collection.*

Cardboard Fire and Marine Insurance advertising fan. 4.5" H x 3.5" L. c. 1950s. $10. *From the author's collection.*

Sections 03.046 through 03.056 Figurines and Sculptures

This subsection presents images of elephant figurines proper. These elephants are models of the real thing and are not like other categories in which the items have some utilitarian purpose like a cigar cutter or a planter. This section contains the largest number of examples simply because "pure elephant" figures are the category most collected by elephant lovers. I know many elephant collectors who have no other categories of elephants; but it would be rare to find a collector of other elephant things who did not also own an elephant figurine.

Before elephants found their way into many of the other types of collectible categories, perhaps the earliest elephant things (besides, perhaps, Woolly Mammoth and Mastodon cave drawings at the dawn of humanity) were figurines created around 5000 B.C. in Egypt. There are certainly examples of Chinese Han and T'ang Dynasty elephant figurines from approximately 207 B.C.-906 A.D.

Since there are so many materials from which elephant figurines are made, I subcategorized these further, to make the location of specific pachys easier. Note that although ivory elephants could have been placed here, they have a category all their own.

Section 03.046
Advertising and Commemorative Figurines and Sculptures

This subsection presents images of non-paper elephants that were made to advertise a company or commemorate an event. Some important commemorative subsets have categories of their own, GOP/Political or Disney for example. I know of the 1961 Sun Rubber elephant as well.

Bronze figurine commemorating President Ronald Reagan. Created by Douglas Van Howd of Sierra Sculpture Inc., an American artist known throughout the world for his bronze sculptures of wildlife and fine oil paintings. The Reagan Edition Elephant is an original sculpture that took Van Howd over one year to complete. Cast in the highest quality bronze and detailed with extreme accuracy showing the lifelike skin, every bone and muscle of the great African Elephant. This is the only piece of artwork personally signed (in the wax in 1980) and numbered by Ronald Reagan. Its insignia reads "Congratulations to a great artist. Ronald Reagan." One of 50 original sculptures, representing each state of the union. 18" H x 17" L. 1980. $50,000 in aftermarket; original price was $5800. *Image Courtesy of Deborah and Lock Lee, and, for extra information, thanks to Holly Thomasson.*

The Elephant and Castle Restaurant, Victoria, Canada, porcelain figurine with howdah. 6" H x 6" L. 1990s. $7-10. *Courtesy of Adele Verkamp.*

Metzeler blue rubber advertising figurine. Metzeler, now a motorcycle tires company, started as a rubber-products company in 1863 in Munich, Germany. A larger example exists, as well as an ashtray. The company's website has elephants on it, including one in a circle as a logo. 2.5" H x 2" L. $5. *From the author's collection.*

Brass souvenir figurine from Lake Placid, New York. 2.5" H x 1.75" L. $5. *From the author's collection.*

Vitalic Products Rubber advertising figurine. 2.25" H x 4" L. c.1920s-30s. $5-10. *From the author's collection.*

Solid copper figurine with blanket on back reading: "Great Seal of the State of Arizona - 1912 Globe-Miami." 1.5" H x 2" L. 1912. $60-75. *From the author's collection.*

Cold cast bronze figurine by the Franklin Mint issued in a limited edition titled: "Giant of the African Plains." It was created by Robert Glen, one of the world's foremost animal sculptors, for the East African Wildlife Society. 7.5" H x 13" L. 1983. $160-200. Franklin Mint also issued a 1978 piece modeled by Neal Deaton, on the "world's largest" Smithsonian elephant; it is 3.75" H x 7" L and is valued at $50-75. *Courtesy of Don Bingham.*

Plastic elephant figurine used to advertise the Fremlins Brewery. c. 1950s. $50. *Courtesy of Mary Marwick.*

Advertising figurine for the Independent Stove Co of Owosso, Michigan, in business from 1906-1953. It was later called Renown Stove Company, and then, Renown Corporation. These elephants were painted in a variety of ways including all silver and all black. One side reads: RENOWN/UNDER-FEED/STOVES;" other side reads: "INDEPENDENT/STOVE CO./ OWOSSO/MICH." 2" H x 3.5" L. 1920s-30s. $60-100. *Courtesy of Julie.*

Painted cast iron figurine commemorating 50th anniversary of the Crane Company. 2" H x 3.75" L. 1905. $40-50. *From the author's collection.*

Bronze figurine commemorating the 100th performance of the DeWolf Hopper Opera Company at the Broadway Theater in New York. 3" H x 4" L. 1891. $50. *From the author's collection.*

Painted pot metal or brass figure commemorating the 1932 World's Fair. 1932. $25. *From the author's collection.*

Section 03.047 Glass and Crystal Figurines and Sculptures

This subsection presents images of glass and crystal elephant figurines. Some of the first glass and crystal makers include Emille Galle (mid-to-late 1800s), who created French Cameo glass vases with elephant images on them. Famous Art Deco glass elephant artists include the Daum brothers, who started a glassworks in 1875 in Nancy, France, Sabino, and Renee Lalique (1860-1945). Angelo and Ercole Barovier in Murano, Italy in the late-1800s to mid-1900s, made interesting elephants. Other famous glass/crystal elephant makers include: Kosta Boda, Orrefors, Bertil Vallien, Archimede Seguso (an artist making glass for Murano, Italy-area makers), Linstrand of Sweden, Bimini (1950s cocktail glasses with embedded elephants), Baccarat, Steuben, and Swarovski, famous for several sizes of crystal elephants and other animals.

Glass figurine from Murano island of Italy, where glassmakers have been making glass since 1291. This elephant is made by the glassworks Vetreria TFZ. Other sought-after Murano glass elephants include those by artist Seguso. 9" H x 6.5 L. 1999. $115-150. *From the author's collection.*

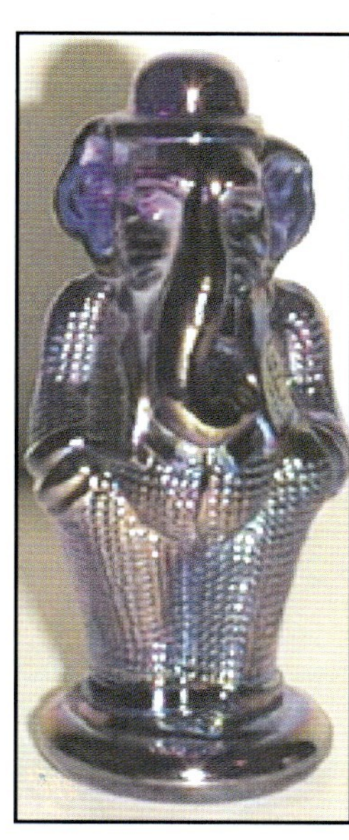

Cobalt blue carnival glass "clown" elephant figurine in tuxedo. Marking on back: "Ron 12." Made by Summit Art Glass Co. in Ohio, these were made in 13 colors after an original mold made in 1918 by Cambridge Glass Co., which is worth $275-350. 4.75" H. $25-30. *Information courtesy of Randy Johnson.*

Lenox lead crystal "Crystal Dancer" elephant glass figurine from Lenox of Germany. 7" H x 5" L. 1999. $120-150. *From the author's collection.*

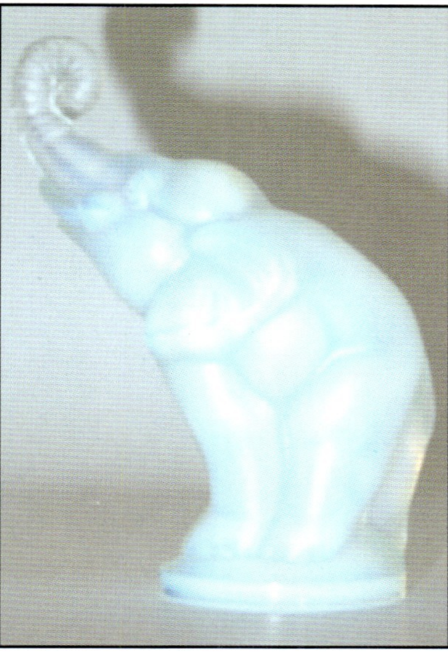

Signed "Sabino-Paris" opalescent glass figurine. Sabino made glass items in the 1920s and 1930s and also from the 1960s to present; the later figurines have a different color than the earlier items. 4.5" H x 2" L. 1990s. $125. *From the author's collection.*

Pele's glass figurine from Hawaii, glass of volcanic origin. 4.25" H x 2.25" L. 1985. $25. *From the author's collection.*

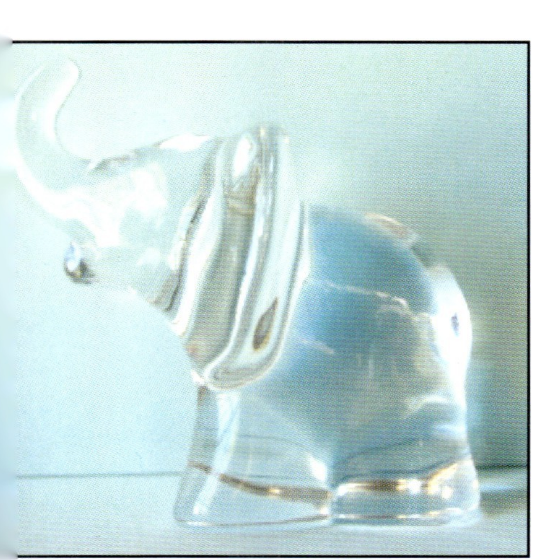

Stozle Kristoll Austria 24% lead crystal. 3.25" H x 3.5" L. 1990s. $30. *Courtesy of Adele Verkamp.*

Swarovski lead faceted crystal figurine; one of several Swarovski makes in different sizes. 2.5" H x 2.5" L. early 1990s. $65-85. *Courtesy of Ted Capell.*

Villeroy and Boch crystal figurine. 3.5" H x 3.5" L. 1990s. $30-40. *Courtesy of Adele Verkamp.*

Rocking Princess House 24% lead crystal figurine. 3.25" H x 2.5" L. 1990s. $30. *Courtesy of Adele Verkamp.*

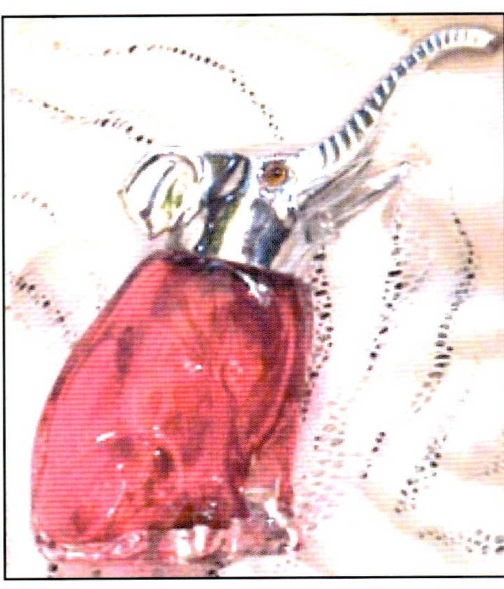

Cranberry glass muffineer. 4.5" H. $20-30.

Signed Fenton vaseline opalescent glass figurine. 3.5" H x 3.5" L. 1999. $25. *From the author's collection.*

Lalique crystal figurine on base with reference #"11801" on the base. 6" H x 6.5" L. 1950s. $350-880. *Courtesy of Jonny.* (E2D4).

A Steuben crystal figurine, shape #8128, designed by James Houston. c. 1964. $700. *Courtesy of www.thievesmarketantiques.com.*

Glass elephant figurine from Loetz, which was founded in 1840 in Bohemia. Loetz closed just after WWII and is as sought after as Tiffany and Galle. Loetz is known most for developing a process for producing glass, after 1898, with a deep blue or gold metallic luster. It was developed by Max Ritter von Spaun the founder Johann Loetz's grandson. 8.5" H x 12" L. $5000. *Courtesy of Alessandro Pron.*

Colored glass elephant figurine from Daum of Nancy, France (making glassworks since 1870s). 7.875" H; smaller size also 5.375" H. $600-800. *Courtesy of Aunt Marla.* (E2D4).

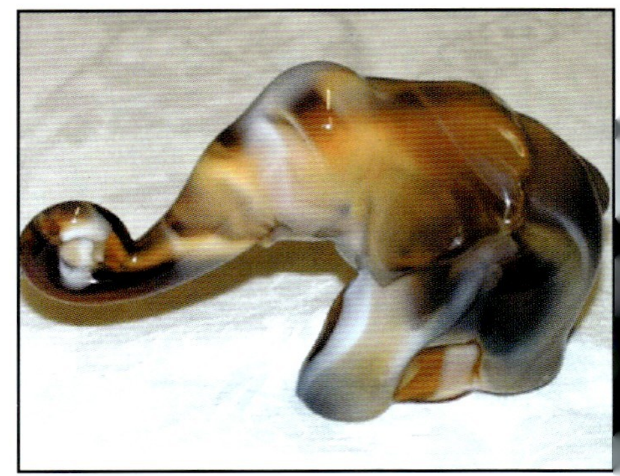

Slag glossy glass, figurine from the Heisey mold, by Imperial of Bellaire, Ohio (which made Heisey mold-based items from 1960-1983. When Arthur R. Lorch became the owner of the Imperial Company the mark became" ALIG" replacing the Heisey mark of "H" after 1968.). 4" H x 6.5" L. c. 1970s. $50-75. *Courtesy of Jim Davidson.*

This glass elephant figurine is part of a series Ercole Barovier designed for Artistica Vetieria e Soffiera Barovier Seguso Ferro. It has the original sticker with number "387/312." Refer to (Barovier and Dougato, 40), and (Heiremans, 143). 8.75" H. 1933. $2500. *Courtesy of Jane Benson. (E2D4).*

Lalique glass elephant called Java. 7" H. 1990s. $2300-3800. Lalique makes other current elephants: Timore, Timori, Timora, all about 3.5" H and Sumatra, which is 7" H. *Courtesy of Harry Fray*. (E2D4).

Cut crystal elephant figurine from Swarovski, part of the Inspiration Africa series. 4.75" H. 1993. $1000-1200. *Courtesy of Holly Ridge Antiques.*

Mosser, of Cambridge, Ohio, cobalt glass figurine. 3.25" H x 3.5" L. 2000. $15-20. *From the author's collection.*

Lavender ice-colored glass figurine from Heisey Club of America, number 335 of 450 in a Limited Edition. HCA mark and "D" for Dalzell Viking, the manufacturer. 5" H x 7" L. 1993. $75. *From the author's collection.*

Boyd Indian Orange and Plum Slag glass pair of figurines; there are 55 colors in the series (up to 2000) of this mold called Zack, made since 1981. Boyd's Crystal Art Glass was formed in 1978 in Cambridge, Ohio. The company was formerly Degenhart Glass. The Degenhart trademark of a heart and a D was replaced with the new Boyd trademark, the diamond B. 4.5" H x 4" L. 1985 and 1988 respectively, $25 each. *Courtesy of Adele Verkamp.*

Section 03.048
Metal Figurines and Sculptures

This subsection presents images of elephants made from different metals and alloys. Douglas Congdon-Martin dates the first iron usage to about 4000 B.C. and cast figures to 1600 B.C. The earliest dated animals are two lions from 502 AD, so I would expect iron elephants to have been made about that time as well. The modern cast iron industry was started by Abraham Darby circa 1700. Congdon-Martin says doorstops are a favorite of cast iron collectors - and there are several cast ion elephant doorstops, including the well-known works of Bradley and Hubbard. Other metal elephants include bookends, bottle openers, doorknockers, and banks.

Figurines have been made from virtually every metal and many alloys. Besides cast iron, probably the most popular metal from which elephants are made is bronze. Forrest (1988) states that the earliest bronzes were made about 3000 B.C. in Babylonia, and the first animals were made in 1500 B.C. Lost wax is the most popular method for making bronze sculptures. The most famous "animiliar" sculptor in bronze is Antoine Louis Barye who produced his first animal in 1831 in France. His famous "Elephant Du Senegal" c. 1870, is one of four by Barye. Another famous artist was Rembrandt Bugatti, who produced abstract animal forms in the early 1900s, including the Bugatti Royale elephant hood ornament. Other bronze elephant sculptures include: "Indian On Elephant" by Charles Valton c. 1880, Roger Godchaux's "Mahout on an Elephant," c. 1900, Carl Akeley's "Wounded Comrade" in the late 1800s, and various artists during the Meiji Period in Japan. One popular motif in these early bronzes (and other mediums) is two tigers attacking an elephant. More recent bronze elephant artists include: Donald Riggs, Loet Vanderveen, Gary Price, and Jack Bryant.

Pewter miniature figurines by Spoontiques: #PP720 High Chair, #HP722 Teeter Totter, #PP7212 Slide. Spoontiques makes several elephants. 2" H x 2" L. c.1990s. $7-15 each. *Courtesy of Adele Verkamp.*

Large patinated lost-wax bronze 3-headed "Erewan" elephant figurine with Hindu God Shiva as rider. Gold trim. 23" H x 14" L. c.1940s. $100. *From the author's collection.*

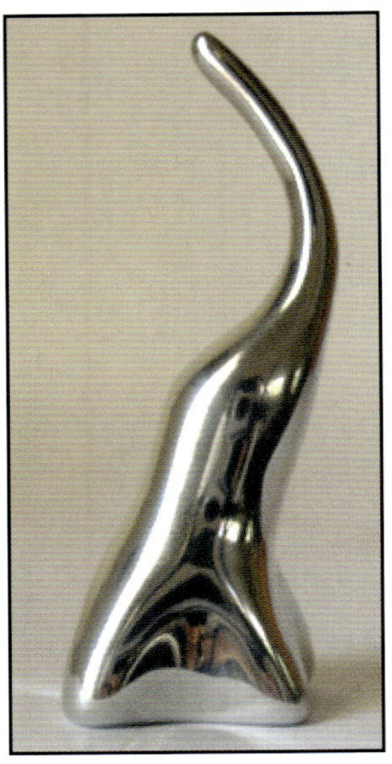

Aluminum figurine marked "181" on bottom. 3.25" H x 1.25" L. c.1990s. $25. *Courtesy of Adele Verkamp.*

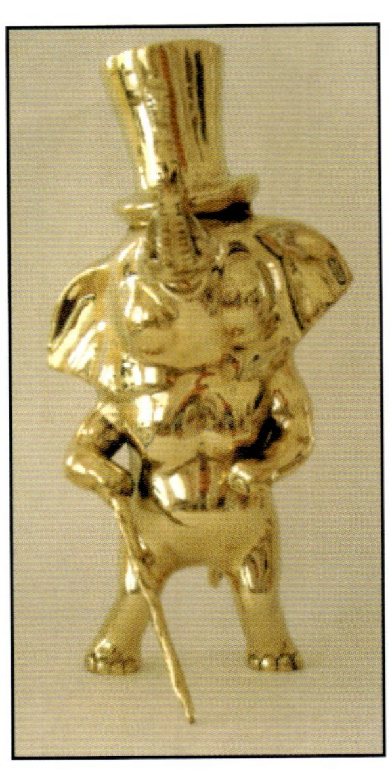

Brass standing elephant figurine with top hat and cane. 24" H. c.1980s. $100. *From the author's collection.*

Bronze reproduction figurine of Antoine Barye's "Jumping Elephant" with marble base and plaque. 27" H x 16" L. 1990s. $800-1200. *Courtesy of Darin Koutney of Bronzes Online.*

Large caparisoned and colored gunmetal figurine made in India. 18" H x 30" L. c.1990s. $300. *From the author's collection.*

Pewter figurine with great proportions and detail. 4" H x 6" L. c.1950s. $35. *Courtesy of Ted Capell.*

Britains Cococubs Tiny Tusks lead elephant figurine; one of several styles. These novelties were given away with Bourneville Cocoa in the 1930s. 1.8" H x 2.1" L. c.1930s. $25. *Courtesy of Toy and Model Collectors Market NZ.*

Brass 14k gold-plated figurine. 1.75" H x 2" L. c.1990s. $50. *Courtesy of Ted Capell.*

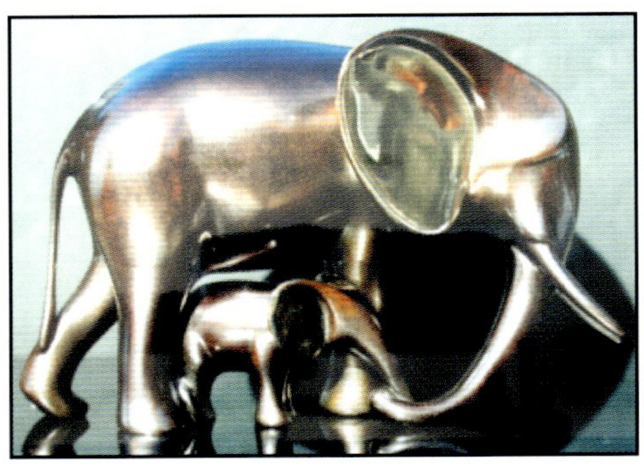

Patinated bronze "Elephant and baby" figurine in an edition of 750 by Loet Vanderveen. 7" H x 10" L. c.1990s. $1175. *Courtesy of Loet Vanderveen.*

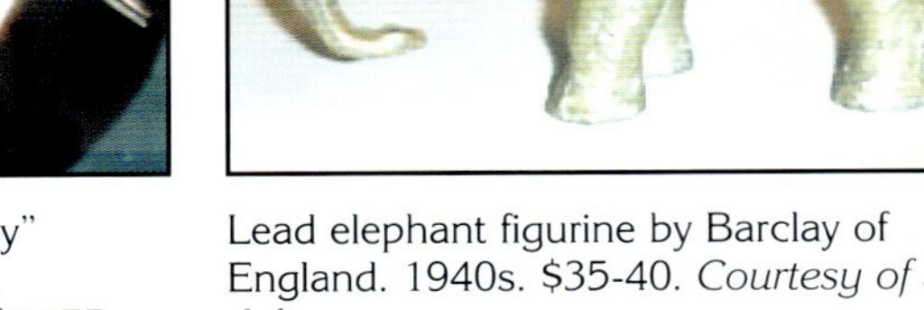

Lead elephant figurine by Barclay of England. 1940s. $35-40. *Courtesy of John Johnson.*

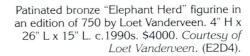

Patinated bronze "Elephant Herd" figurine in an edition of 750 by Loet Vanderveen. 4" H x 26" L x 15" L. c.1990s. $4000. *Courtesy of Loet Vanderveen.* (E2D4).

Patinated bronze "Triumphant" figurine by noted artist Gary Price. 8" H. c.1990s. $500. Three larger sizes exist: 12", 24" and 37" high. *Courtesy of Legacy Gallery of Scottsdale, Arizona and Gary Price.* (E2D4).

Copper-patinated brass figurine of elephant on a ball. 13" H x 5" L. $25. *From the author's collection.*

Green-patinated brass figurine made in California. 30" H x 16" L. c.1999. $175-225. *From the author's collection.*

Gilded bronze standing elephant figurine balances on his tail. 6" H x 3.25" L. c. early 1900s. $75. *From the author's collection.*

Metal Alymer figurine made in Spain, in HO scale, titled "Elephants of Carthage," on a pedestal. 1" H. c.1980s. $25. *Courtesy of Ric Bracamontes.*

Chrome-plated solid metal figurine. 4.5" H x 4" L. c.1990s. $40. *Courtesy of Adele Verkamp.*

Dansk Designs figurine, solid zinc and plated with brass. 2.5" H x 2.25" L. c.1990s. $25. *Courtesy of Adele Verkamp.*

Austrian bronze elephant figurine is hand-highlighted, for which the Austrians were noted. The piece weighs nearly 5 pounds and has the original ivory tusks and pads on its feet. 5" H x 9.5" L. c.1940s. $400. *Courtesy of Joseph M. Wright.* (E2D4).

Older bronze figurine made with lost wax process with removable rider probably India or SE Asia. 5" H x 6.25" L. c.1800s. $175. *From the author's collection.*

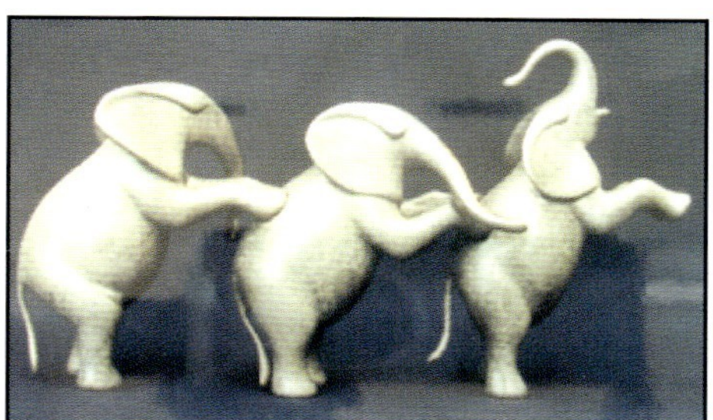

Patinated bronze "Standing Elephants" figurine by famed artist, Vanderveen. Loet produced many other elephant works. 7" H x 12.5" L. c.1990s. $1435. *Courtesy of Loet Vanderveen. (E2D4).*

Very large patinated bronze "Large Elephant" sculpture by noted jewelry creator and wildlife artist Donald Riggs. This elephant is made in a limited edition of 25. This elephant and Riggs's other works in general (including several other elephants) have cubist and art deco influences, yet retain flowing lines. 36" H x 12" W x 54" L. c.1990s. $15,000. *Courtesy of Donald Riggs*. (E2D4).

Art Deco bronze elephant figurine by Karl Hagenauer of Austria. 3" H. 1930s. $350.

Hudson pewter circus elephant figurine. Signed "J. Wanat." Base bottom has the following: "2477, (c) 1982 Hudson Fine Pewter, U.S.A." Circus elephant with a bright red star on the covering on the elephant's back. 2.75" H x 1" L. 1982. $25-50. *Courtesy of Joanne Bulczak of American Collectibles*.

Bronze figurine made using lost wax process, with elaborate howdah; part of set of three from Orissa (in NE India). 7.75" H x 6" L. 1910-1920. $100. *From the author's collection*.

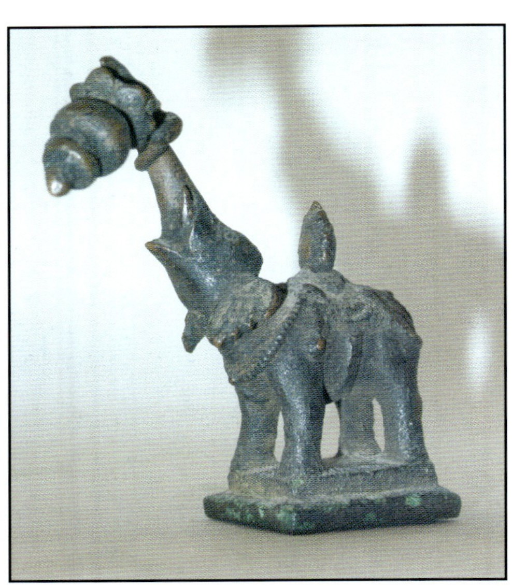

Chola dynasty Indian bronze figurine of Gajendra, the lord of the elephants, with verdigris patina. It has lotus flower petaled torch in trunk and flame aurealla on back and is situated on a stepped base. Serpent head tail holders and bead form trappings. 3" H x 4" L. c.10th-12th century. $350. *From the author's collection*.

Great bronze sculpture of an running elephant, called "Elephant du Senegal." It is by the well-known French sculptor, and father of the Animalier School, Antoine-Louis Barye. Barye was born 1796 in Paris and died in 1875. You find his works in the Louvre in Paris and in other great museums. The sculpture is signed "Barye" with the foundry mark of "F. Barbedienne. Fondeur." It also has a gold-plated mark "FB" by the same foundry. 5.2" H x 7.2" L. $7500. There are many reproductions bringing in the range of $1500-5000. *Courtesy of Ulrich Wagner*. (E2D4).

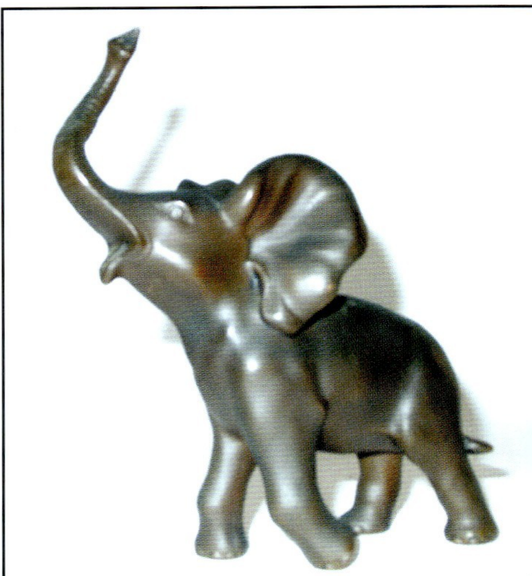

Solid copper figurine. 6.75" H x 5.5" L. c.1950s. $25-50. *From the author's collection*.

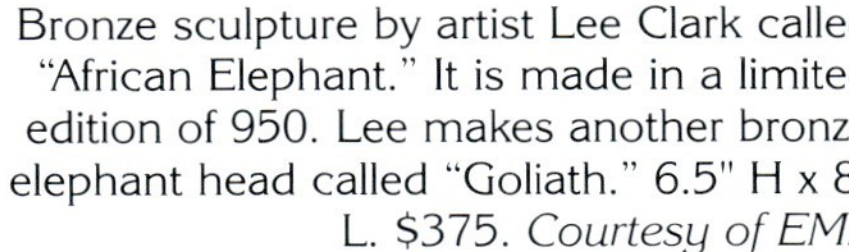

Bronze sculpture by artist Lee Clark called "African Elephant." It is made in a limited edition of 950. Lee makes another bronze elephant head called "Goliath." 6.5" H x 8" L. $375. *Courtesy of EMI*.

Flat pewter elephant figurine from Belgium or France. 2.5" H x 3" L. 1991. $25. *From the author's collection.*

Detailed small bronze figurine by Theo B. Starr Inc., of New York. 1.75" H x 2.25" L. c.1900-1930. $25-50. *From the author's collection.*

Brass wire over brass figurine inlaid with turquoise and coral stones. 1.75" H x 3" L. c. early 1900s. $50. *From the author's collection.*

Meiji period signed Japanese bronze elephant fighting a tiger attack; this is a classic design motif. The elephant has ivory tusks and the tigers have inset glass eyes and relief striping. The figurine is standing on the original gnarled root wood base. 1880s. 18" H x 21" L. $1500-2000. *Courtesy of ted_pappas@westcoast-estates.com.* (E2D4).

Bronze elephant figurine signed "Carl Sorensen" on the side of the base. The bottom is also signed with the number "159" and other undecipherable characters. Carl made Arts and Crafts metalwork at the turn of the century in Philadelphia, Pennsylvania. 7.5" H with base. c. 1910-1920. $350.

Pewter cast sculpture marked "Wilton Columbia U.S.A." (Pennsylvania). 4" H x 5.75" L. $25. *From the author's collection.*

Section 03.049 Mineral Figurines and Sculptures

This subsection presents images of elephants made from natural minerals. There are so many types of minerals that one could easily spend a lifetime getting one example elephant of each type and still never be sure one has all types. It is also not clear to me that every mineral and rock variety has been used by a carver to make an elephant - but one could have a lot of fun finding out! If you want to know more about a particular mineral, the National Audubon Society makes excellent field reference guides to rocks and minerals (published by Knopf). A note about valuations of elephants sculpted or carved from the precious and semi-precious minerals (e.g., jade, malachite, lapis, ruby, coral etc.): the value of the elephant increases non-linearly with respect to the size. In other words, where a 2" high malachite elephant might sell for $50, a 6" high one would be $500 - not thrice as much, but 10 times as much!

Carved rhodonite figurine. 3" H x 4.25" L. c.1990s. $25. *From the author's collection.*

Carved chalk (limestone) figurine. 3" H x 4.5" L. c.1990s. $15. *From the author's collection.*

Carved soapstone figurine from India with baby carved inside hollow mother. 3" H x 4" L. c.1990s. $10. *From the author's collection.*

Carved marble figurine from Mexico. 9" H x 10" L. c.1980s. $50. *From the author's collection.*

Large, carved Burmese jade figurine from the C'hing Dynasty. 4.6" H x 7" L. c.1800s. $200-300. *From the author's collection.*

Carved hematite figurine, an iron-based mineral. 2" H x 2" L. c.1990s. $25. *From the author's collection.*

Carved soapstone figurine from Africa. 6" H x 8" L. c.1980s. $35. *From the author's collection.*

Carved coal figurine from Pennsylvania. 7" H x 7" L. c.1970s. $25. *From the author's collection.*

Honan jade inlay over wood frame figurine with semi-precious stone inlays and 24kt gold filigree blanket and chains. 15" H x 20" L. c.1980s. $500-700. *From the author's collection.*

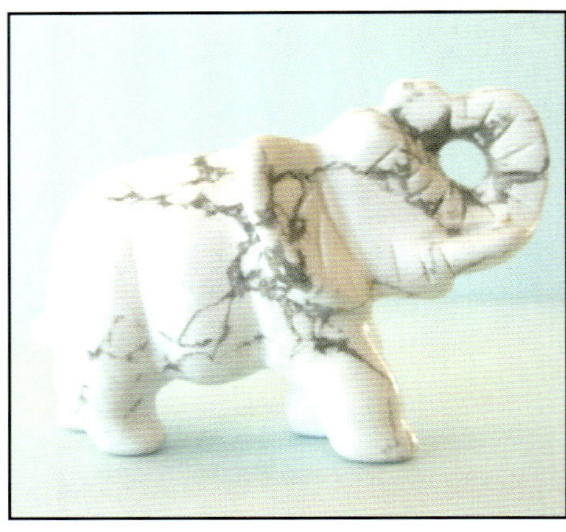

Carved howlite figurine. 2" H x 2" L. c.1990s. $10-15. *From the author's collection.*

Carved lapis figurine, a semi-precious stone. 1.75" H x 2" L. c.1980s. $35-50. *Courtesy of Ted Capell.* (E2D4).

Carved malachite figurine, a semi-precious stone. 1.5" H x 2.25" L. c.1990s. $35-50. *Courtesy of Adele Verkamp.* (E2D4).

Carved marble figurine. 5.5" H x 7" L. c.1990s. $35-50. *From the author's collection.*

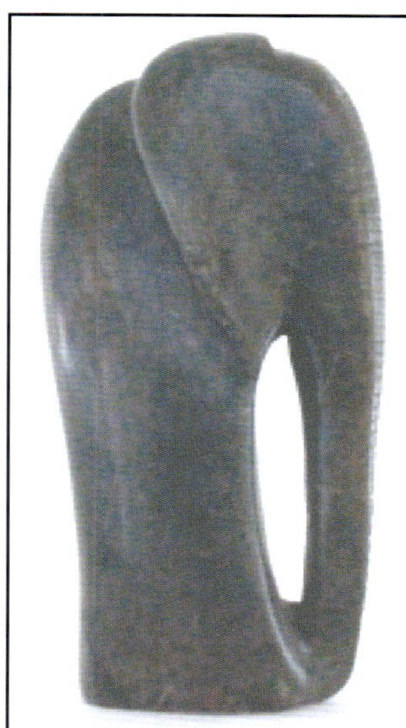

Carved green serpentine figurine from Africa. 5" H x 3.75" L. c.1990s. $50-75. *Courtesy of Adele Verkamp.*

Carved black onyx figurine. 3.5" H x 2" L. c.1980s. $10. *From the author's collection.*

Black marble carving from Zimbabwe. 5" H x 2" L. 1998. $30. *Courtesy of Janet Wojciechowski.*

Blue green obsidian figurine, also called crystal obsidian; it occurs on top of the usual black obsidian. This stone is from mainland China. c.1990s. $20. *Courtesy of Rex Silfant.*

Carved Picasso jasper figurine. 3.25" H x 4" L. c.1990s. $20. *From the author's collection.*

Indiana limestone sculpture used as prototype for a series of bronze water fountain castings by artist Meg White. It weighs approximately 3,000 lbs! 51" H x 54" L. 1990s. $24,000; the bronze fountains are $28,000. *Courtesy of Meg White.*

Carved and painted chalk figurines marked ABCO (Alexander Backer Co.). 4" H and 3.25" H. c. late 1940s. $15 set. *Courtesy of Jane Chapman.*

Carved chrysoprase figurine. 3" H x 3.5" L. c.1990s. $100. *From the author's collection.*

Carved leopard skin jasper figurine. 2" H x 2" L. c.1990s. $25. *From the author's collection.*

The ultimate elephant sculpture! An extremely rare black coral sculpture by noted artist Bernard K. Passman. Passman has been using black coral, found only in the Caribbean, since 1974, after he first visited the island of Grand Cayman. He has produced commissioned black coral works for Prince Charles and Princess Diana, HRH Queen Elizabeth, and Pope John Paul II. He is also a prominent jewelry designer, using black coral in his designs. His sculptures of elephants followed a trip to Africa where he studied the elephant's form and he says that they are his favorite sculptures! This sculpture is actually three pieces, the two base pieces and the log with the large elephant on it bridging them. The left base has two baby elephants following. 14" H x 27" L. 2000. $250,000. *Courtesy of Bernard K. Passman Galleries.* (E2D4).

Indian agate carving from Hong Kong. 2" H x 2.25" L. 1990s. $100. *Courtesy of Dan Ryder.*

Opalstone family of elephants from the Shona of Zimbabwe. Signed by Lloyd Mellusi. 14" H x 35" L. 1990s. $1500-2000.

White meerschaum figurine. The purest, most easily worked Meerschaum, a type of clay, is mined in Eskisehir, Turkey. In nature it exists within other stones as a round deposit. It is very light, floats on water when first mined and is easily worked, becoming hardened after time. It has been called "White Gold" due to this hardening process. Meerschaum was found by a Hungarian in the 17th century and is mined in a number of mines around the Eskisehir area. 5" H. 1990s. $35-50. *Courtesy of Merchants of the Globe.*

Matrix opal elephant figurine comes from Magdalena Jalisco, one of the most prolific and famous opal mines in Mexico. Carved by Alvaro Cuellar who has gained international recognition for his amazing craftsmanship. 1" H x 1.5" L. 1990s. $10. *Courtesy of Lisa Topolniski.*

Stone figurine from Africa. 4" H x 6" L. 1998. $40. *Courtesy of Janet Wojciechowski.*

Figurine made of marble and inlaid with semi-precious stones including lapis lazuli, malachite, turquoise, onyx, etc. It was handcrafted in Agra, India using the traditional method of inlaid stones, as featured on the Taj Mahal Pietra dura. The inlaying of semi-precious stones into marble was brought from Italy to India under the rule of Shah Jahan and is used in Muslim religious art. 3" H. 1990s. $75. *Courtesy of Julian Heinz of Colors of the Orient, Inc.*

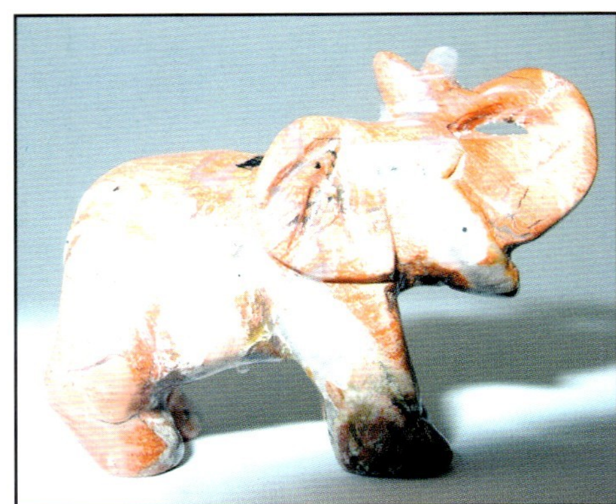

Carved brecciated red jasper figurine. 1.5" H x 2" L. c.1990s. $15. *From the author's collection.*

Carved rainbow florite figurine. .75" H x 1.5" L. c.1990s. $10. *From the author's collection.*

Relatively rare red amber figurine from Africa. 7" H x 10" L. 1870. $1800-2000. *Courtesy of Robert C. Amann Sr.*

Carved zoistite figurine with ruby inclusions. 3" H x 5" L. c.1990s. $150. *From the author's collection.*

Section 03.050 Miscellaneous Figurines and Sculptures

This subsection presents images of elephant figurines whose composition does not fit into one of the other main subcategories of materials, or constitutes a mix of materials.

Smokey quartz figurine. 5" H x 7" L. 1990s. $165.

Carved snowflake obsidian figurine. 2.25" H x 2.75" L. c.1990s. $25. *From the author's collection.*

Carved tiger eye figurine from Hong Kong. 2" H x 2.75" L. c.1990s. $25. *From the author's collection.*

White jade figurine from China. 7" H x 9" L. 1990s. $100-160. *Courtesy of Dan Ryder.*

Figurine made of carved fishbone pieces on wood frame with brass and semi-precious stone trim on wood stand. 18" L x 14" H. 1990s. $50-100. *From the author's collection.*

Figurine of brown leather over wood frame. 12" L x 12" H. 1970s. $35. *Courtesy of Ted Capell.*

Carved buffalo horn figurine from India. 3.5" H x 2" L. 1980s. $10. *From the author's collection.*

Wood figurine covered with cloth then a clear coat of epoxy or shellac; from China. 5" L x 7.25" H. 1990s. $25. *Courtesy of Ted Capell.*

Clay whimsical figurine by Dave Grossman Designs of St. Louis Missouri. 4" H x 4.5" L. c. 1970s. $15. *Courtesy of Adele Verkamp.*

Clay whimsical figurine from Israel. 4.5" H x 4.5" L. 1998. $25. *Courtesy of Adele Verkamp*.

Figurine carved from coconut. 8" H x 6" L. 1980s. $20. *From the author's collection*.

An antique paper mache candy container made in Germany. The pointed tusks are 1" long and are made of white glass. The toenails are painted a darker gray. There is a mark on the head sleeve that looks like "Germany." 3.75" H x 6" L. 1885-1920. $200-300. *Courtesy of Robert Stubbs*.

Figurine of hammered bronze sheeting over solid wood frame. This elephant has bells hanging free from its feet and sides, with a larger bell dangling around its neck. Made in India but no tourist trademark present. 10" H x 12.5" L. c.1930s. $200-300.

Indian folk art figurines are 100% cotton cloth with glass facets over a cotton batt interior. 9" H and 7" H. c. 1960s. $20 and $15. *From the author's collection*.

Cow bone figurine. The bone is granulated and mixed with epoxy and poured into a mold. Made in Mexico. 6" L x 7.25" H. 1990s. $20. *Courtesy of Ted Capell*.

Section 03.051
Plaster Figurines and Sculptures

This subsection presents images of elephants made of plaster, which is nothing more than lime or gypsum mixed with sand and water. The slip is poured into an elephant mold and hardens. Plaster elephants are easy enough to make for the home hobbyist as long as he/she can find or make appropriate elephant molds.

Detailed plaster elephant mom and baby with accurate painting. 13" H x 12" L. c.1990s. $50. *From the author's collection*.

Colorful plaster figurine with "Versil Statuary '70" embossed on one side. This design is very similar to one by Arnels in the Porcelain Figurines category. 5.75" H x 9" L. 1970. $25. *From the author's collection*.

Plaster Elephant "Buddha" in full meditative squat. 36" H x 21" L. 1980s. $275. *Courtesy of Ted Capell.*

Plaster shelf sitter by Hassman. 7" H x 4" L. 1996. $25. *Courtesy of Adele Verkamp.*

Section 03.052 Plastic, Celluloid and Rubber Figurines and Sculptures

This subsection presents images of plastic, celluloid, and rubber elephant figurines. Plastic is one of the most common materials from which to fashion just about anything, and elephant collectibles are no exception. Breyer is one of the major manufacturers of plastic animal figurines. Especially noted for their horses, they made several elephants too. Be careful when evaluating elephants that look like ivory or Bakelite, which is an early, and now sought-after (read relatively expensive) plastic; they could be suitably colored inexpensive plastics or celluloid!

Plastic figurine with detailed blanket and trim. 3.5" H x 4.5" L. 1970s. $25. *Courtesy of Ted Capell.*

Molded hard plastic elephant. 6" H x 9" L. c.1970s. $20. *From the author's collection.*

Molded plastic elephant with colored blanket by Bergen Toy and Novelty USA. 2.5" H x 3.5" L. c.1970s. $10. *From the author's collection.*

Solid plastic woolly mammoth. 1:24 scale from Bullyland Inc.; commissioned for the Stuttgart Museum of Natural History. 5" H x 9" L. 2000. $20-30. *From the author's collection.*

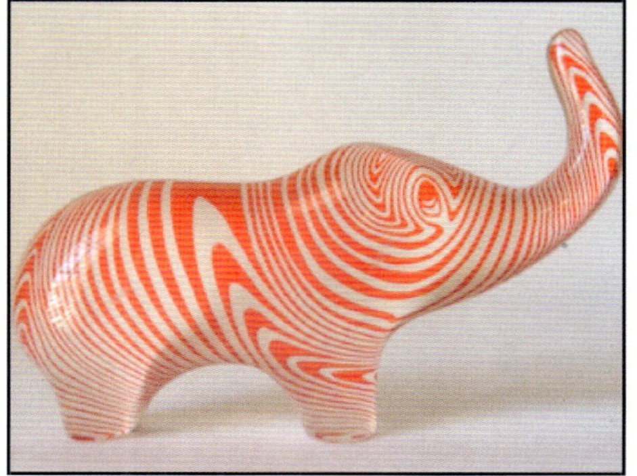

Lucite with colored swirls embedded in the mold. 4" H x 6.5" L. c.1960s. $25. *Courtesy of Adele Verkamp.*

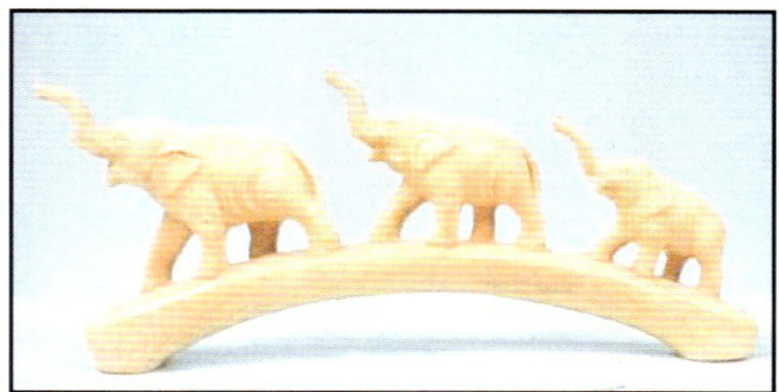

Celluloid over plaster ("chalkware") bridge with 3 elephants. 2.375" L. 1940s. $10. *Courtesy of Jane Chapman.*

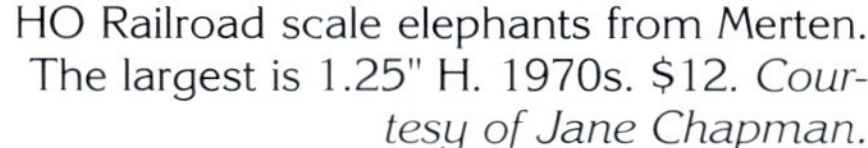

HO Railroad scale elephants from Merten. The largest is 1.25" H. 1970s. $12. *Courtesy of Jane Chapman.*

Plastic gray figurine, Mold #91 from noted animal model maker Breyer. 6.5" H x 7.5" L. c.1950s. $30. *Courtesy of Ted Capell.*

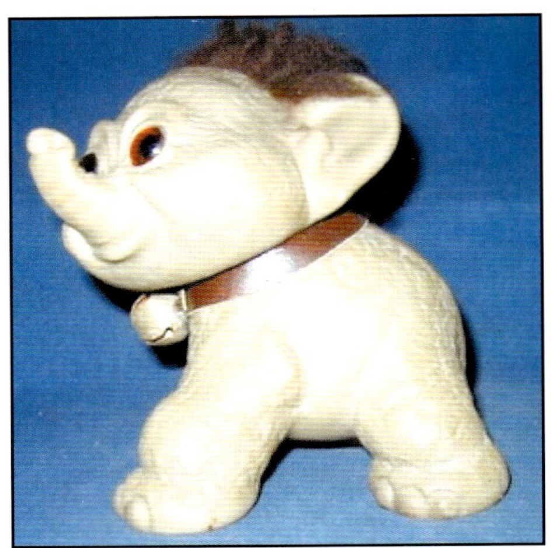

Dam troll made by The Troll Company situated in the northern part of Denmark. Marked: "Made in Denmark, Dam Patent" on feet. Textured squeezable vinyl skin. Head is jointed and neck has original collar with bell. 5.5" H x 4.5" L. c.1950s. $125-150.

Rubber squeezable elephant that squeaks, by Edward Mobley Company. 10.5" H x 9" L. 1950s. $25. *Courtesy of Tom Angelotti.*

Section 03.053
Porcelain, Pottery, China, and Ceramic Figurines and Sculptures

This subsection presents images of "pottery" elephants, pottery being a broad term for ceramics, porcelain, earthenware, etc. Michael Schneider (1990) has details on the making of pottery, which is categorized mostly according to the firing temperature and materials used. Earthenware is the softest pottery, fired at 1470-1830F. Redware, yellow ware, majolica, faience, terra cotta and most ceramic pieces are earthenware. Stoneware is fired at 2280 to 2370F. Porcelain can be soft paste or hard paste variety. Soft paste was used by Europeans a few hundred years ago while experimenting and is fired at 2100 degrees F. Hard paste is fired at 2550F and is commonly referred to as porcelain. Bone china is made at 2240 with bone ash added to the paste. Firing at higher temperatures means the elephant is more durable and more costly, all other things being equal. Of course, desirability, rarity, and the manufacturer's reputation for quality are the predominant determinations of price in this category.

Limoges porcelain is one moniker for quality porcelain. It is so named because, in the mid-18th century, the formula for hard paste porcelain was discovered and used in the Limoges region of France. The clay in the area had kaolin and feldspar and this combination led to the first hard paste Limoges porcelain, which does not craze, produced in 1771. Since early porcelain, from the first 100 years or so until about 1860, is rather rare and expensive, most items, including elephants that you are likely to see and collect were made since about 1860. Many of the vases made in the Limoges area, for example, by Pouyat, have elephant head handles. Haviland also produced some elephants. Staffordshire is another general location in England that had many potteries that produced elephant figurines and spill vases, shown here and in the Vases and Amphora category.

Clay figurine by California Creations. 3.175" H x 3.25" L. $15. *From the author's collection.*

Teeter and Totter from The Herd by Martha Carey. 3.5" H x 7.5" L. 1990s. $60. *Courtesy of Adele Verkamp.*

Porcelain figurine signed by artist Jay Lims. 4.5" H x 8.25" L. $25. *From the author's collection.*

Clay figurine made in Peru by Casals. 6" H x 5.5" L. 1990s. $15. *Courtesy of Ted Capell.*

Porcelain figurine by Princeton Galleries, a division of Lenox in Malaysia. 4" H x 6" L. 1990s. $15. *Courtesy of Adele Verkamp.*

White porcelain figurine by Lomonosov of St. Petersburg, Russia. 4" H x 5.5" L. 1990s. $25. *Courtesy of Adele Verkamp.*

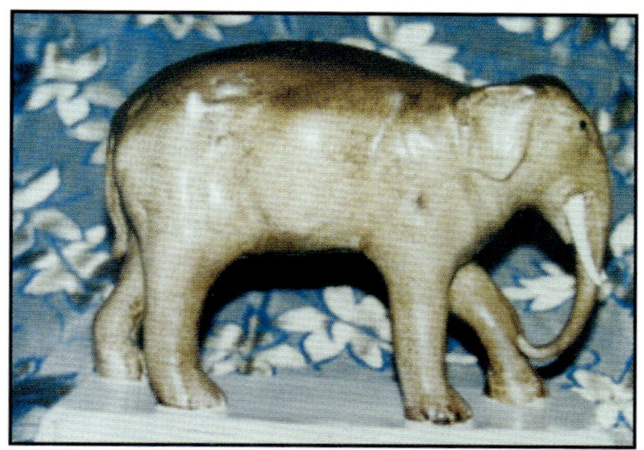

Porcelain figurine from a Staffordshire pottery, on base, unmarked. Some figures like this have had "JUMBO" applied in gilt. 8.25" H x 10" L. c.1850. $800-1200. *Courtesy of Paul Morgans and Jeff Gregory.* (E2D4).

Mosser porcelain figurine. 4" H x 3.5" L. 1990s. $25. *Courtesy of Adele Verkamp.*

Cloisonné figurine, which is enamel between brass (or other metallic) bands serving to divide an enamel (usually) surface. 3" H x 3" L. 1990s. $35. *Courtesy of Adele Verkamp.*

Crinkled ceramic carousel elephant titled Possible Dreams with number "6591813" marked on bottom. 4.5" H x 4.5" L. 1998. $35. *Courtesy of Adele Verkamp.*

Hand-painted ceramic figurine with "Lakpigan Made In Sri Lanka" inkstamp on bottom. 2.5" H x 3.25" L. $15. *From the author's collection.*

Glazed ceramic figurine by California Originals of Torrance, California. 6" H x 9" L. 1990s. $50. *Courtesy of Adele Verkamp.*

White porcelain figurine by Homco of Mexico. 3" H x 5" L. 1993. $10. *Courtesy of Adele Verkamp.*

Pottery with inset tiles on head, characteristic of maker Artesania Rinconada of Uruguay, who has been making earthenware ceramic material with enamel glazes since 1972. Each figurine is unique by virtue of different tiles. It is marked with the usual "aR" on the bottom of one of his feet. This is model Elephant 08 from the Classic Collection. 3.5" H x 4" L. $50. *Courtesy of Adele Verkamp.*

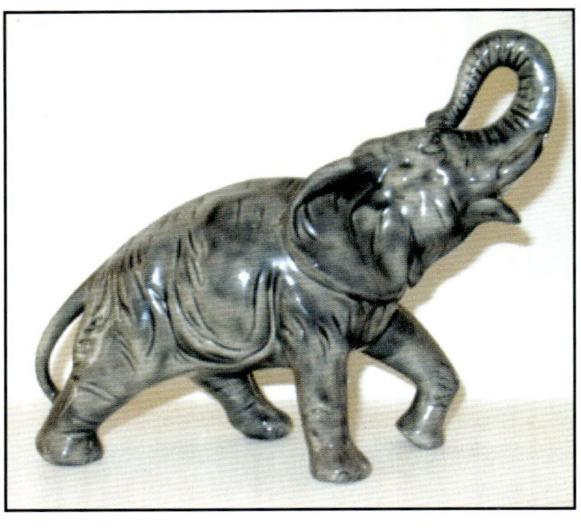

Ceramic figurine by Claytime Ceramic. 9" H x 11" L. 1980s. $20. *From the author's collection.*

Clay figurine from Beasties of the Kingdom series by Raya Designs of Orlando, Florida. 2.5" H x 3" L. $10. *From the author's collection.*

Porcelain figurine signed "Elephant" in script on bottom; made in Japan sticker with "UC" line 1 and "CTI" line 2. 6.25" H x 8" L. 1980s. $15. *From the author's collection.*

Porcelain figurine from Capidimonte with crystal ball on trunk from Swarovski. 3.5" H x 5" L. 1990s. $50-75.

Clay figurine with "Made In Uruguay" on bottom; possibly by A. Rinconada. 3.5" H x 4.25" L. $30. *From the author's collection.*

Allach porcelain figurine designed and signed by Kaerner with hallmark. 2.75" H x 3" L. c. 1938. $600+. *Courtesy of www.petziman.de.*

Large example of intricate cloisonné figurine. 4" H x 5" L. $400.

Ceramic figurine with fuzzy "fur" by Kunstlerschutz of West Germany; has Handiwork sticker. 2.5" H x 3" L. c 1960s. $15. *From the author's collection.*

Ceramic mother and baby made in Japan, with sticker on bottom: "UC" on line 1 and "CTI" on line 2. 2.75" H x 3.5" L. 1970s. $10. *From the author's collection.*

Porcelain figurine by Princeton Galleries, a division of Lenox in Malaysia. 4" H x 3.5" L. 1990s. $15. *Courtesy of Adele Verkamp.*

Porcelain figurine by AGC Royal Collection. 5" H x 3.5" L. $15. *From the author's collection.*

Porcelain sitting elephant made by Lefton. "HCR 1873" printed on bottom. 3.75" H x 3.25" L. c. 1960s. $10. *From the author's collection.*

Painted pottery figurine from Mexico. 5.75" H x 3.25" L. 1980s. $20. *Courtesy of Adele Verkamp.*

Clay figurine by Russ Berrie with "14212" impressed on bottom; made in China. 3" H x 2.5" L. $10. *From the author's collection.*

Large porcelain elephant carrying flower bouquet, made by Capidimonte of Italy. 15" H x 19" L. 1999s. $125. *From the author's collection.*

Porcelain hand-painted elephant from the Delft region of Holland; also see Tiles category. 2.5" H x 4" L. 1999. $10-15. *From the author's collection.*

Porcelain figurines signed by artist "By Cheri 81." 4.25" H x 3" L and 3" H x 2.5" L. 1981. $10 pair. *From the author's collection.*

Porcelain figurine with plastic-plugged hole in bottom. "Celebrity" printed on bottom with Made In Taiwan Republic of China sticker. 3" H x 4.5" L. $10. *From the author's collection.*

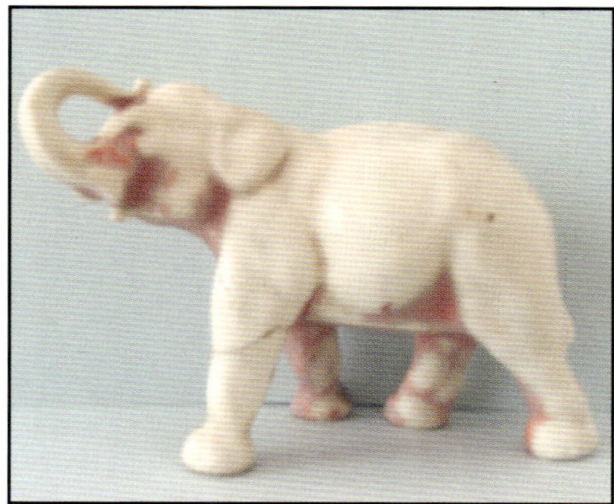

Bisque porcelain figurine with "Erphila Germany" on bottom. 4" H x 6" L. 1940s. $50. *From the author's collection.*

Rare Amphora porcelain figurine. The Amphora factory was located in the Teplitz-Turin area of Bohemia during the late nineteenth and early twentieth centuries, according to *Kovel's Antiques and Collectibles Guide of 1999*. This figurine has "1034" impressed on stomach and 2 gold lines on foot. 2.875" H x 4" L. 1890-1910. $300-500. *Courtesy of Terry Wurden.* (E2D4).

Colorful painted ceramic figurine by Arnels. 6" H x 9" L. 1990s. $25. *From the author's collection.*

Ceramic Arts Studio, who made items from 1941 to 1955 in Wisconsin, produced these Tembo and baby Tembino figurines. 2.5" H x 4.75" L and 5" H x 5" L. 1950s. $80-100 and $40-60. *Courtesy of Joseph M. Wright and Lynn Garrard.*

Made of Parian china, this is the largest elephant in the Floral Treasures Collection made by Belleek Pottery Works Co. of Ireland. The blanket is highly embossed and in each corner of the blanket there is an applied flower and leaves in mother-of-pearl. 6" H x 8" L. 2000. $125-175. *Courtesy of Kelly.*

Porcelain elephant by Bretby of England. 10" H x 16" L. early 1900s. $250. *Courtesy of Ed Buchanan.*

Pottery figurines, a Josef Original Mother and Baby, in separate pieces. 6.5" H and 3.375" H. c.1980s. $45 pair. *Courtesy of Jane Chapman.*

Porcelain elephant riding cart by Beachstone/Renaker of Brazil, formed by Mary Renaker. It was in business from 1974 until about 1990. $15. *Courtesy of Jane Chapman.*

Large porcelain elephant with hand-enameled accents, from Lenox's Palace Collection; others include a baby sitting elephant. 8" H x 8.5" L. c.1990s. $125. *From the author's collection.*

Bone china Coalport figurine on base with inkmark; Coalport was purchased by Wedgwood after this model was made. 5" H x 5" L. late 1980s-early 1990s. $50. *Courtesy of Collect-US of Eynon, Pennsylvania.*

Ceramic elephant "Angelings" with wings by Creative View. 6" H x 4" L. 1994. $10. *From the author's collection.*

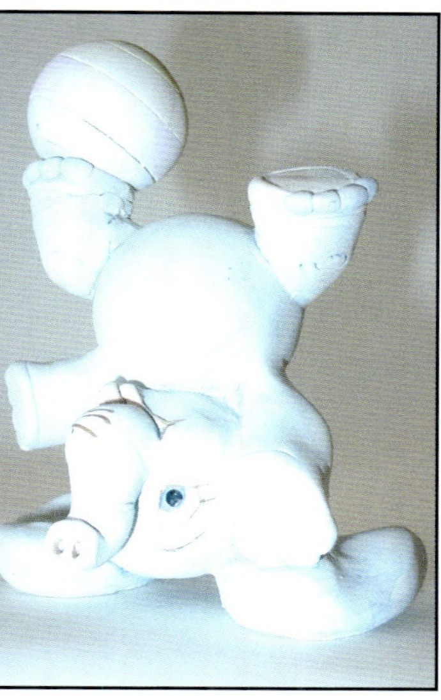

Clay "Balancing Act" elephant standing on front legs by Dreamsicles, #10584. Signed by artist: "Kristin 98." Made in Mexico by Cast Art Industries. 4.25" H x 3" L. 1998. $15. *From the author's collection.*

Freeman-McFarlin porcelain elephant standing on hind legs. 1.75" H x 1" L. 1950s. $20. *Courtesy of Jane Chapman.*

Kreiss ceramic mammoth with cave boy, featured in *Aikens, 1999.* 4.75" H x 6.25" L. 1950s. $60-80. *Courtesy of Pat and Larry Aikins.*

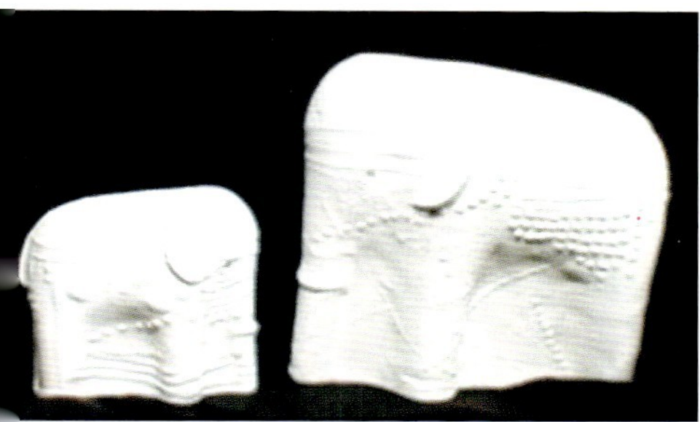

These are large and small ceramic figurines from the Terra Series that artist Bertil Vallien did for Rorstand of Sweden. 7.5" H x 8" L and 3.75" H x 4" L. 1998. $75 and $50. *From the author's collection.*

Porcelain elephant with baby by Goebel of Germany. 7" H. $25-40.

Standing porcelain baby elephant by McFarlin Potteries. 5" H x 7.5" L. late 1970s-1986. $35. There is also a sitting baby and it is 4.5" x 5.75". *Courtesy of Jane Chapman.*

Dancing matte glaze ivory porcelain elephant is from Caliente Pottery by Harold Johnson. It is marked on the bottom of his feet. 4.5" H. c.1940s. $25.

This is a one of a kind, Rosalene elephant, hand-painted with purple and cream blossoms by Marilyn Wagner, a well-known artist for Fenton. The cream blossoms are fretted with crystal glass. 22K gold highlights the floral designs, the tip of his nose and his mouth. 3.75" H x 4" L. $100.

Herend baby elephant with lavender fishnet design from Herend of Hungary. This is model #15429. 3" H. 2000. $75. *Courtesy of Laszlo Alfoldy.*

Porcelain from Gilde Porzellan of Germany with "Wertnolle echie Handerbeil" on bottom. 3.25" H x 4" L. 1980s. $10. *From the author's collection.*

Lladro of Spain (started in 1951) made this large elephant family, model #1150G, as well as several other elephant figurines. 15" H x 16" L. 1971. $800-1100. A similar model #2297M, in matte finish, was issued in 1995. A Mother and Baby figurine, model #1151 has a similar gray glaze as the Family shown and was also made in 1971; it is valued at $400-500. Note that NAO elephants are also made by Lladro. *From the author's collection.*

An elephant on base with turquoise glaze from Rookwood Pottery. The piece is marked with the RP and flames mark, mold no. 6498 and a scratched 64. This piece is from the end of Rookwood's production when only a limited number of molded pieces were being made. Production had been shifted from Ohio to Mississippi in 1961 to reduce costs and the pottery finally closed in 1967. 3.75" H x 4.5" L. 1964. $150. *Courtesy of Peter Bergstrom.*

Hagen-Renaker elephant standing on circus platform model #A880. 2.375" H. 1987-1989. $20. *Courtesy of Jane Chapman.*

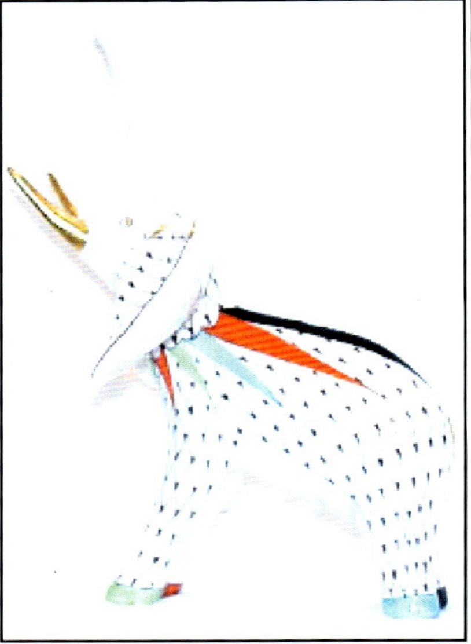

Hand painted porcelain "Garden" elephant with fishnet design from Hollohaza Porcelangyar RT of Hungary. 8.5" H x 6" L. 1999. $80. *Courtesy of Hollohaza USA, a division of Crystal Dreams Inc.*

Lenci (founded in 1919 in Turin, Italy) porcelain nymphet on elephant. 16" H x 13.5" L. 1930s. $4500-6500. *Courtesy of Michael Verker.* (E2D4).

China, Lusterware figurine, impressed Japan. 2.375" H and 1.5" H. c.1940s. $10 and $5. *Courtesy of Jane Chapman.*

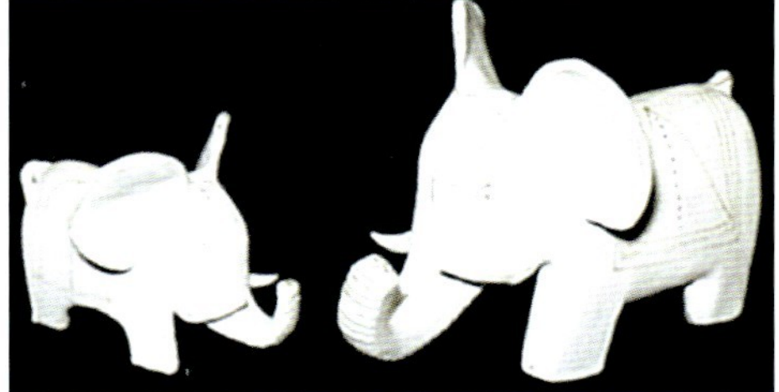

This pair of elephants was designed by Guido Gambone and each is signed "Gambone Italy." 8" H x 12" L and 5.25" H x 8" L. c.1950s. $400-500 *Courtesy of Terry.*

A miniature porcelain figurine by Meissen. Meissen figurines are usually very realistic, even though this one seems out of proportion. This was modeled by Meissen artist Kaendler in the mid-1700s. The Meissen mark of the crossed swords is on his belly, with impressed characters "Y86," and "1568." 1.5" H x 3" L. c. 1970-1980s. $225. *Courtesy of Morris Peterson.*

Porcelain elephant by Kaiser Porcelain in Germany; an open edition. 4.75" H x 6.25" L. Original mold was made in 1956. $150-200. *Courtesy of Robert Arnado.*

Porcelain with characteristic hand-painted (in blue) fishnet design from Herend of Hungary. This is model #15266. Herend makes several models ranging from a large 15" L model for about $1500, to several smaller ones ranging from $75-150; they also made some plain white elephants. 3.5" H x 3.5" L. 1990s. $100-125. *From the author's collection.*

Lladro porcelain figurine called "Hindu Children riding elephant;" model #5352. 9" H. 1986. $450-500. There is a similar model called" Hindu Children," model #2298, but with brighter colors, issued in 1995, retired in 2000, with a height of 9.5" and a similar value. *Courtesy of Greg Monroe.* (E2D4).

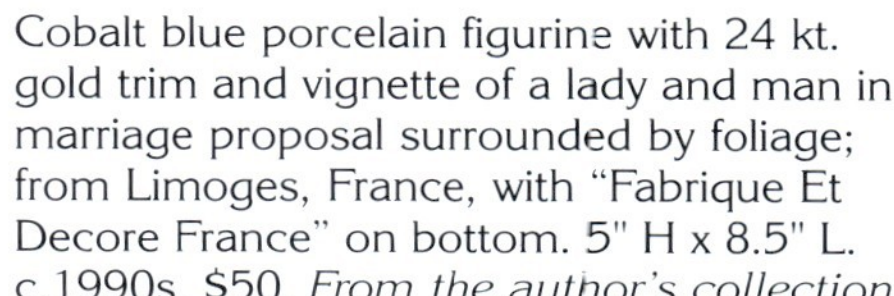

Two ceramic elephants on a bench by artist Lila Stewart in the Wild Things Series. 7" H x 9" L. c.1990s. $50. *From the author's collection.*

Cobalt blue porcelain figurine with 24 kt. gold trim and vignette of a lady and man in marriage proposal surrounded by foliage; from Limoges, France, with "Fabrique Et Decore France" on bottom. 5" H x 8.5" L. c.1990s. $50. *From the author's collection.*

Ceramic figurine with green Japan ink stamp. 3.5" L. c.1930s-40s. $5. *Courtesy of Jane Chapman.*

Porcelain elephant by Gebruder Heuback of Germany, founded in 1820 and known for its porcelain dolls. c. 1930s-40s. $500.

Small porcelain elephant made for Royal Crown Derby. Both the large (which has a Howdah) and small elephant, made in 1990, are decorated in the old Imari pattern. Other decorations on the head, legs, etc. are interpretations of traditional Indian ornaments. 4.5" H x 6.25" L. 1990. $325; the large size is $500-600. Other RCD elephants include the following: Harrods Large Elephant Limited edition of 150 made for Harrods of London (1999), Mulberry Hall Baby Elephant in a limited edition for Mulberry Hall of York (1997/98/99), and a Solid Gold Band Baby Elephant in a limited edition of 1000 produced for Gumps of San Francisco (1990/1999).

Two porcelain figurine/spill vases from Staffordshire potteries of England. There were hundreds of potteries in the Staffordshire area of England from the 1700s onwards; refer to *Godden 1975* for more information on Staffordshire pottery. c. mid-1800s. $800-2000 ea. *Courtesy of Ally Godfroid.*

Porcelain figurine by Loza Electrica of Mexico. 1.25" H x 3" L. c.1990s. $10. *Courtesy of Jane Chapman.*

Black bone china figurine with colored flow on back with label: "Fait Au Canada Made, Poterie Evangeline Pottery Inc., Labelle, Quebec." 4.5" H x 7" L. 1960s. $20. *From the author's collection.*

Santa is riding atop an elephant with an elf as his safari guide; an intricate sculpture of hand-painted, cold-cast porcelain by Maruri Studio in a limited edition of 5,000. 9" H x 12" L. 1996. $225. *Courtesy of Dave Nash.*

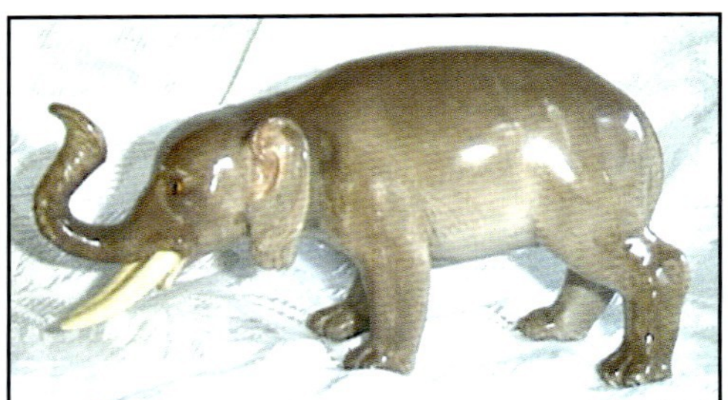

Porcelain Meissen figurine. 3" H x 6" L. from the Marcoloni period, late 18th or early 19th century. $200. *Courtesy of Jennifer Konzalski.* (E2D4).

A Chester Nicodemus pottery elephant with a great drip glaze. This piece is unmarked like most of his miniatures, but is pictured in *Riebel, 24.* 1.25" H x 2" L. $75. *Courtesy of Scott Shook.*

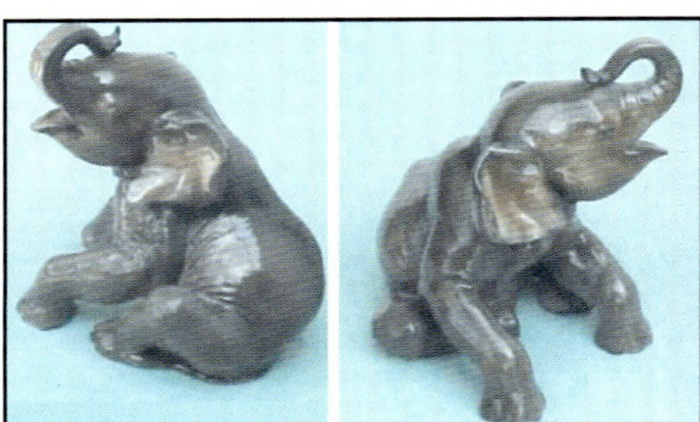

Sitting elephant by Rosenthal, a German firm making pottery since 1880. 6.375" H x 5" L. 1940s. $200-225. *Courtesy of John Morrison.*

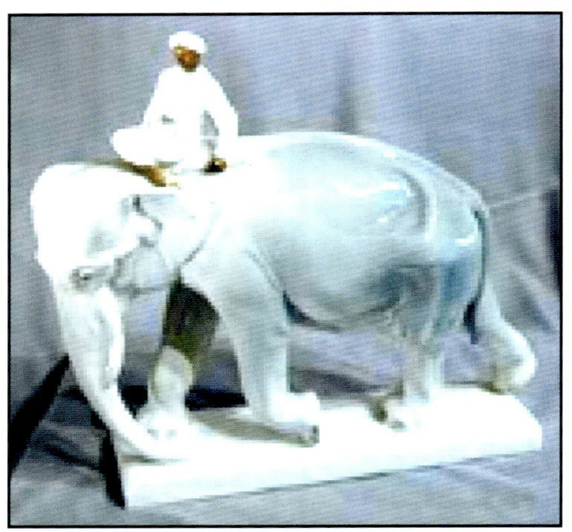

Porcelain elephant and rider created by well-known Meissen artist Walther. The figurine has the Meissen crossed swords mark. Size 11.5" H x 12" L. 1903. $2000-2500. *Courtesy of The Russian Heritage, New York, New York.* (E2D4).

Metlox Dumbo figurine similar to a model made by American Pottery. 5.5" H 1950s-60s. $50-75.

Royal Doulton white bone china figurine with flowers in mini-planter on back. 3.25" H x 4" L. 1999. $15. *From the author's collection.*

Pair of ceramic elephants by Rio Hondo. 2.25" and 1.25" H. 1930s-1940s. $10 pair. *Courtesy of Jane Chapman.*

Matched pair of porcelain elephants with a "Norleans Japan" gold and black silver label. 6" H x 8" L. c. 1950s-60s. $25-40. *Courtesy of Herman Kalfen.*

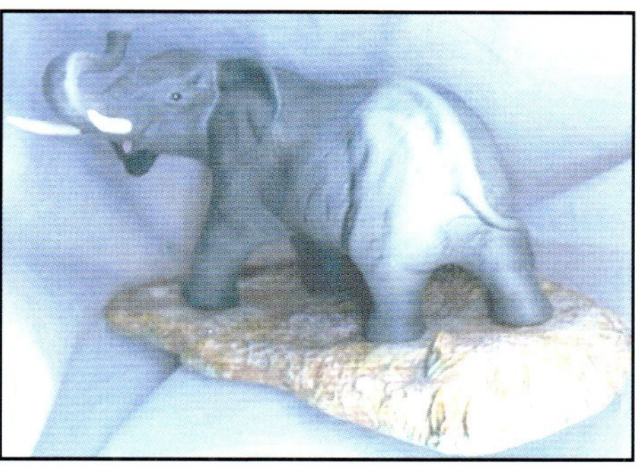

Porcelain elephant in the Royal Doulton Endangered Species series (*Royal Doulton Standard Catalogue of Royal Doulton Animals, Jean Dale, Second Edition*). 6" H. 1991-92. $100-125. *Courtesy of Paul Ruddle.*

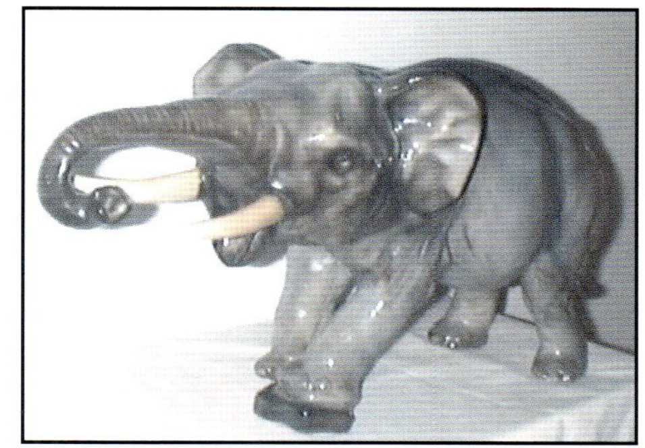

Royal Doulton porcelain elephant model #HN 2640, designed by Charles Noke, a noted Doulton designer. It is from the Prestige series and is titled "A Fighter Elephant with White Tusks." 12" H x 24" L. 1952-1992. $1500-2000. (E2D4).

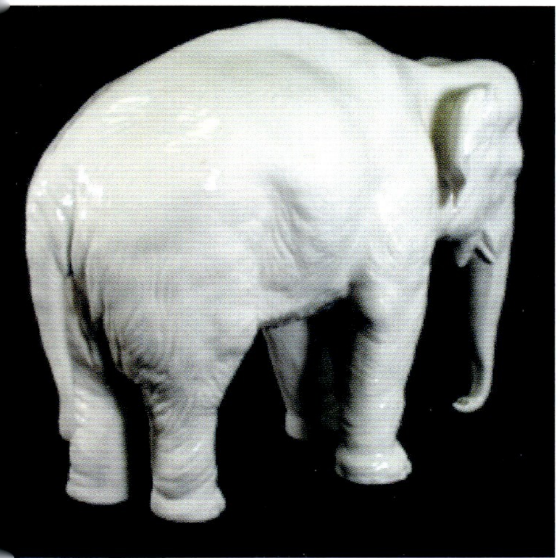

Large porcelain Indian elephant from one of the finest porcelain producers in Germany, Nymphenburg of Bavaria (current name is Royal Nymphenburg) started in 1753. This elephant was modeled by Neuhauser and Karner. Recent reproductions have a contemporary shield mark. 10" H x 12" L. 1920s. $800-1200. *Courtesy of Bruce Kodner.* (E2D4).

DeLee (or possibly Walker) porcelain figurine. 2.5" H. x 2.5" L. late 1940s. $20. *Courtesy of Jane Chapman.*

A miniature Staffordshire porcelain elephant, dark brown with white tusks, standing on a white base decorated with green foliage. It can also be seen in *Harding 2000, p. 266.* 3" H x 3" L. 1800s. $250-350. *Courtesy of Ally Godfroid.*

Large gray porcelain Asian elephant. He has ivory-colored tusks and is made by Lenox, a company started in 1906. 6" H x 9.5" L. 1996 (1999 Ivory colored). $60. This elephant is similar to the Lenox bone-colored model called Classic Majestic Elephant, which is valued at $100-150. Lenox currently has about 30 elephant models available.

Raku pottery figurine from African Express of South Africa. Raku is attributed to Zen Buddhist Monks of 16th Century Japan and can be translated as "enjoyment of freedom," according to Joyce E. Furney who runs a pottery studio. This Raku elephant was baked in a kiln for 12 hours, then glazed and fired to 950 degrees C. The elephant was then placed in a smoldering bed of sawdust to create the pattern on the glaze. 4" H x 6" L. 1999. $60. *From the author's collection.*

Pottery elephant from Italian maker Raymoor. 1950s-60s. $50-75.

Ceramic "Netter" figurine by Rosenthal Netter Inc. in Italy. 4" H x 6" L. c.1960s. $15. *From the author's collection.*

Royal Doulton flambe bull elephant; Charles Noke is the designer. 5.25" H x 8" L. 1926-1962. $200-225. Royal Doulton also made a smaller and larger version; the latter is model 4532, is 7" H x 10" L, and ranges in value to $1000. *From the author's collection.*

Roselane Pottery (Pasadena, California) ceramic figurine on wood block. 8.25" H x 12.5" L. 1940s. $200. *Courtesy of Lou Blew.*

Alexander, the Cybis porcelain circus elephant with ball from the Circus Collection. There is also a Cybis carousel elephant that ranges to $1400. 7.5" H x 6" L. 1975. $500. *From the author's collection.*

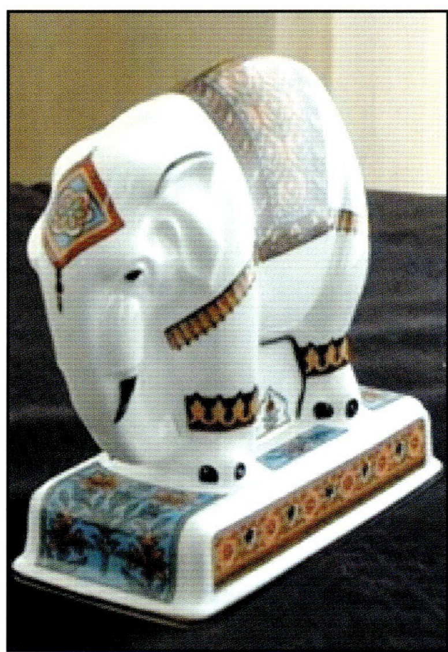

Bone china elephant on base by Wedgwood of England (established 1759); from their Noah's Ark Collection. 5.5" H x 5.5" L. 1993. $150.

Royal Doulton flambe figurine titled "Motherhood" is model #HN 3646. This is from the Images of Fire Collection. 8.5" H. x 9.25" L. c. 1991. $300-400. There is a similar Mother and Young model that is 3.5" H x 5" L and is valued at about $200. (E2D4).

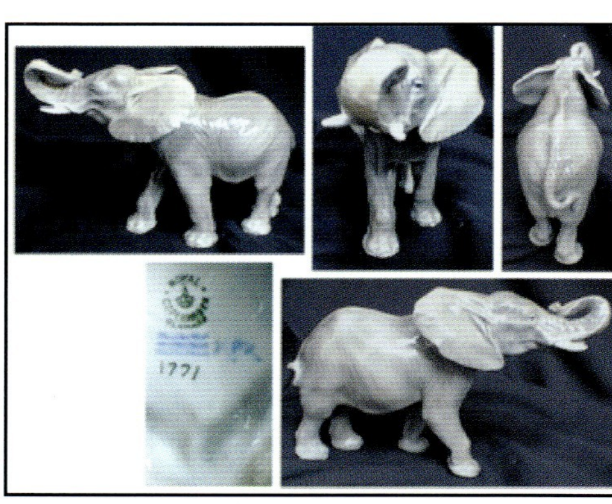

Porcelain figurine from Royal Copenhagen of Denmark (since 1772). Stamped with "1771" and the company mark (crown over 3 wavy lines) on bottom. 6" H x 9.5" L. c. 1950s. $300-500. *Courtesy of Dana Okey.*

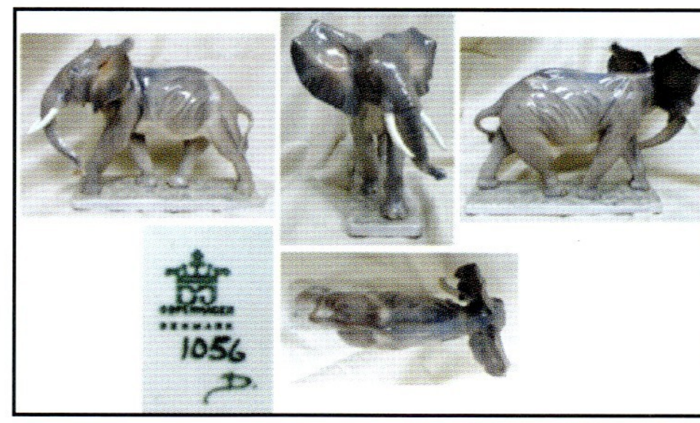

Royal Copenhagen porcelain figurine on stand, stamped with model #1056 and company mark on bottom. 9" H x 12" L. c. 1950s. $300-500. *Courtesy of Dana Okey.*

Royal Dux of Czechoslovakia, making pottery since 1860, created this large porcelain elephant with characteristic triangular sticker reading "Royal Dux Bohemia" and a raised pink triangle reading: "Royal D" with a circle around the "D" on the bottom, along with several numbers. 7" H x 12" L. $75-100. A larger version exists at approximately 10" H x 13" L, and is valued at about $175. There is also a white 10" H elephant valued at $260. *From the author's collection.*

Royal Haeger (in Illinois since 1914, started using Royal Haeger name in 1938) porcelain figurine with baskets hanging by brass holders on each side. 14" H x 21" L. $100. *From the author's collection*.

Porcelain elephant by artist Sascha Brastoff who made pottery in California from 1953 to 1973. $150-250. *Courtesy of Drexel Riley.*

A Van Briggle (making pottery in Colorado since 1901) porcelain elephant in matte Persian Rose color with a spray of blue on his back. His right front foot is marked "VB" and the left front foot has the Van Briggle mark. There is also a Van Briggle paper-weight valued at about $40-75. 5.75" H x 8" L. 1940s. $100-200. *Courtesy of Judith Opfell*. (E2D4).

A figure of Jumbo, P.T. Barnum's famous circus elephant who died in 1885. Made by Parr and Kent in Staffordshire, England. Title in raised capitals reads "JUMBO." It is decorated with majolica glazes and has blue eyes. 11.5" H x 10" L. c.1882. $500-800. *Courtesy of Monique and Patrick Kelley.*

Porcelain elephant made by the Stangl Pottery Co. of Flemington, New Jersey. with original label in his left ear. 5" H x 3.5" L. $50-60; A smaller elephant is valued at $125. *Courtesy of Linda Lewis.*

Large porcelain "amphora" elephant from the Turin-Teplitz area of Bohemia with impressed mark of Crown. 23" H x 26" L. c. 1900. $1000-1800. *Courtesy of Bob.*

Zsolonay (of Hungary since 1862) porcelain elephant with characteristic eosin glaze. 3" H x 4" L. c.1950s. $150-300. Recent versions are about $30. *Courtesy of Sun Glo Antiques*.

Porcelain elephant with jasper tusks and glass eyes by Wedgwood of England. 3.5" H x 5" L. early 1900s. $400-500. *Courtesy of Robert Hamilton.*

Wade (of England) Whimzie series elephant walking with trunk over head. 1.375" H x 2" L. 1983-1985. $10. A 1953-59 model is valued at $50. Wade also made an elephant with circus drum from 1993-1998. Wade Whopper series elephants are larger. *Courtesy of Jane Chapman.*

Porcelain elephant by Zanesville Pottery in Ohio. 6" H x 9.5" L. c.1917. $150-200. *Courtesy of Mike Verker.* (E2D4).

Bisque porcelain figurine with gold accents from Walker-Renaker. 2.875" H x 2.25" L. c.1950s. $10. *Courtesy of Jane Chapman.*

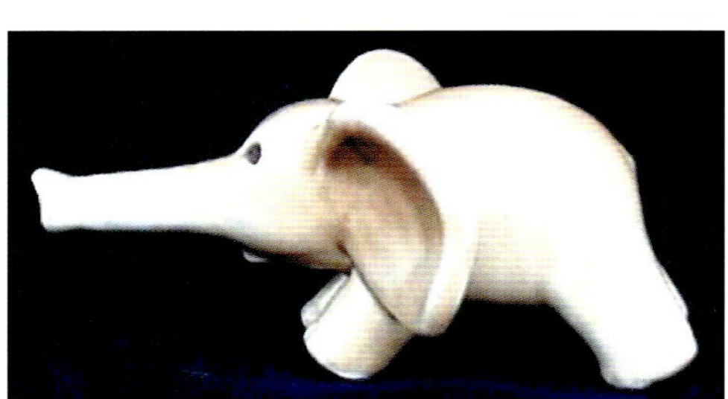

Porcelain elephant produced by SylvaC of England (SylvaC was the trade name for pottery produced by Shaw and Copestake Ltd), with embossed #3140 on his underside and an ink mark. SylvaC also made several models/sizes (#768/69/70/71) of a standing elephant ranging from $40-75 in value. 3.75" H x 4.75" L. c.1950s. $130. *Courtesy of Swan Collectibles, England.*

Section 03.054 Resin and Composition Figurines and Sculptures

This subsection presents images of elephants that are made of resin or some composite material.

Resin elephant head created by Marra Gallery, from original that was carved from bark. 10.5" H x 8.5" L. c.1980s. Value: $75. *From the author's collection.*

Resin figurine by Casinelli. 1990s. $50.

Sitting resin elephant on wood base by Country Artists Limited in Stratford Upon Avon, England. 3" H x 3.25" L. $130. *Courtesy of Adele Verkamp.*

Resin sculpture by John Perry, a noted animalier, on cypress base. 8" H x 12" L. c.1991. $100. *Courtesy of Janet Wojciechowski.*

Resin Hamilton Collection Peanut Pals "Having a Ball" and "A Slice of Fun" by artist Tom Newsom. Made in China. Both 4" H x 3.5" L. 1996. $7.50 each. *Courtesy of Adele Verkamp.*

Resin Hamilton Collection Peanut Pals "Lazy Days" and "Merrily, Merrily" by artist Tom Newsom. Made in China. 2.5" H x 4" L and 4" H x 3.75" L. 1997. $8 each. *Courtesy of Adele Verkamp.*

Resin reclining elephant in underwear by Enesco, made in Republic of China. 3.5" H x 2.25" L. 1970. $8. *From the author's collection.*

Large fiberglass figurine made for a circus display. 4.6' H x 5' L. c.1950s. $500. *Courtesy of Ira Galvish, Art In Antiques, Miami, Florida.*

Resin #812 from Russ and Wallace Berrie. 6.25" H x 3.25" L. 1970. $5. *From the author's collection.*

Resin Hamilton Collection: Peanut Pals - Sculpture Collection "Trick or Treat" and "Spring is Sprung" by artist Michael Adams. Made in China. both 4" H x 3.5" L. 1997. $8 each. *From the author's collection.*

Resin Hamilton Collection: Peanut Pals, Waterfall Ways, and Elephant Days collection "How does your garden grow" and "Catch of the Day," made in China. 2.5" H x 4" L and 4" H x 3.75 L. 1997. $8 each. *From the author's collection.*

Colored resin pair with good detail. 4" H x 4" L. c.1990s. $5. *From the author's collection.*

Colored resin. 5.5" H x 5.25" L. c.1950s. $20. *From the author's collection.*

Resin figurine from series of "Tuskers" elephants created by Barry Price, who once worked for Royal Worcester. This one is called "Sally Shading Baby." 7.5" H. 2000. $45. *Courtesy of Barry Price and George Kamm.*

Section 03.055 Shellaphants

This subsection presents images of elephants composed of different types and sizes of shells. The shells are usually glued onto some elephant-shaped substrate like wood, plaster, or, perhaps, Styrofoam. Sometimes the shells are pressed into soft clay or putty. Other shellaphants are made by gluing the appropriately sized and shaped shells together to make the figurine of an elephant.

Elephant made of different types of shells. 5.5" H x 6" L. c.1990s. $5. *From the author's collection.*

Elephant made of clay with shells glued over surface. 8.5" H x 4" L. c.1990s. $15. *From the author's collection.*

Section 03.056 Wooden Figurines

This subsection presents images of carved wood elephants. Other than pottery, wood is the most popular or common material in which elephants are fashioned. It seems that many manufacturers or artists using wood as their medium do not sign or otherwise identify their works, at least not nearly so much as many other types of elephants. This explains why there are not as many wood figurines in this book as porcelain. Even though there may be as many wood elephants around as porcelain ones; not nearly as many different major manufacturers work in wood, probably due to it not being amenable to mass production techniques; at least that is my theory.

Stained wood figurine from Thailand by APM Company Ltd. 12" H x 13" L. c.1980s. $60. *From the author's collection.*

Teak figurine on stand with ivory tusks. 10" H x 8.5" L. c.1980s. $50-75. *From the author's collection.*

Stained wood elephant on rocker, caparisoned with brass and copper decorations, including a solid brass tail. 7" H x 9.25" L. c.1980s. $25-40. *From the author's collection.*

Large, very heavy, extremely detailed elephant with baby, carved from one piece of ironwood with carved wood tusks. 15" H x 23" L. c.1990s. $350-400. *From the author's collection.*

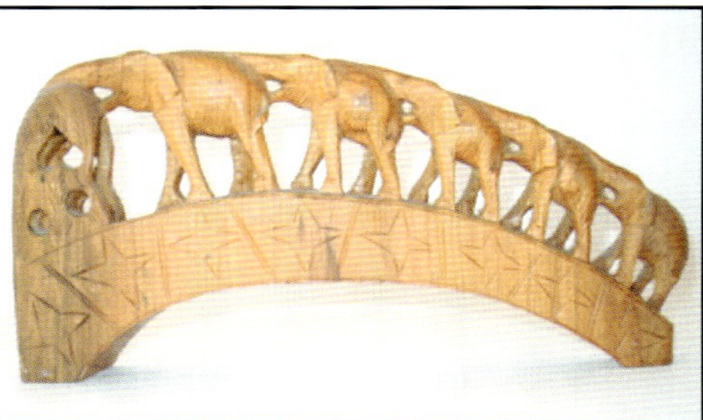

Teak bridge carved in Thailand; a product of Norleans. 7.5" H x 14" L. c.1980s. $25-40. *From the author's collection.*

Carved wood elephant that was part of a beam from a temple in India. 9.25" H x 10.5" L. c.1800s. $175-200. *From the author's collection.*

Carved wood elephant. 4.5" H x 4" L. c.1980s. $25. *From the author's collection.*

Carved, rosewood elephant with ivory tusks and toenails. 12" H x 13" L. c.1990s. $75-100. *From the author's collection.*

Carved, mahogany elephant with ivory tusks with inlaid brass design on back. 6" H x 9" L. c.1990s. $35-50. *From the author's collection.*

Carved, stained wood elephant by Paul Marshall from Thailand. 5.25" H x 6.25" L. c.1980s. $20. *From the author's collection.*

Fat wood elephant with brass tusks. 11" H x 15" L. c.1960s. $75. *Courtesy of Ted Capell.*

Rosewood elephant family. 11" H x 4" L. c. early 1990s. $55. *Courtesy of Ted Capell.*

Teak elephant pulling trunk on base; carved in Thailand. 6" H x 12" L. c. early 1970s. $20. *From the author's collection.*

Ebony bridge with inlaid ivory toenails and eyes. 5" H x 14" L. $40-50. *Courtesy of Janet Wojciechowski.*

Painted wood elephant purchased in Chinatown, San Francisco. 5" H x 4" L. c.1991. $15. *Courtesy of Janet Wojciechowski.*

Tall accentuated rosewood elephant. 10" H x 15" L. c.1970s. $35. *Courtesy of Ted Capell.*

Kaddham wood elephant with howdah, from India. 7" H x 3" L. c.1980s. $20-25. *Courtesy of Ted Capell.*

Painted wood shelf-sitter. 6" H x 3.5" L. c.1990s. $20. *Courtesy of Adele Verkamp.*

Wood elephant stack. 23" H x 3.75" L. c.1990s. $75. *Courtesy of Adele Verkamp.*

Large carved wooden head. 19" H x 23" L. c.1990s. $175. *Courtesy of Ted Capell.*

Ironwood elephant and baby with beautiful grain, carved in Burma. 8" H x 10" L. c.1960s. $175. *From the author's collection.*

Rosewood elephant made in Kenya. 4.25" H x 3.25" L. c.1970s. $25. *From the author's collection.*

Rosewood with ivory tusks and toenails and inlaid eyes from India; one of a pair. c.1950s. $100. *Courtesy of Janet Wojciechowski.*

Schoenhut articulated wood elephant, strung with string, with a rope tail and leather ears. The tip of the trunk is leather or rubber 7.5" H x 10" L. 1930s. $200-300.

Section 03.057 Fountains

This subsection presents images of some elephant fountains. The one made of bronze is molded from a master carving made of granite, which is shown in the Mineral Figurines section.

Exterior painted solid concrete pool fountains. 24" H x 20" L. 1994. $100 each. *From the author's collection.*

Cast iron fountain. 18" H x 16" L. c. 1920s. $300. *From the author's collection.*

Composition lighted fountain with an elephant family standing on a rock cliff above a waterfalls. 12" H x 14" L. 1990s. $150-200. *Courtesy of Don and Priscilla Chalfant.*

Bronze cast fountain of baby elephant; original sculpture carved in Indiana limestone by Meg White. 51" H x 54" L x 36" D. 1990s. $28,000. *Courtesy of Meg and Don Lawler.*

Section 03.058 Furniture

This subsection presents images of elephant furniture. I placed the pottery elephant stand in this category as well; they are perhaps the most common "furniture" item. I have seen some pretty fantastic furniture with an elephant motif and plan to get many more examples in the next volume.

Recent bronze table base made in Thailand. It is composed of three elephants in the trunk-up position to hold a glass or stone top resting on the tips of the trunks. 19" H x 18" L. base. 16" L. 1990s. $400-500. *Courtesy of Estrada-Jindawed.*

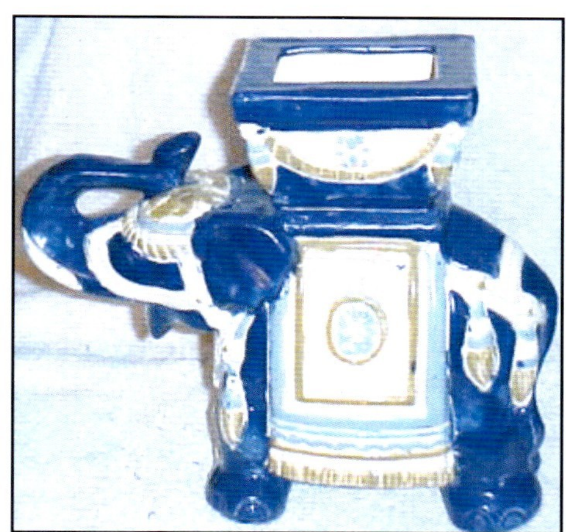

Ceramic stand. 20" H x 24" L. $75-125.

Solid mahogany chair with leather trim. 72" H x 36" W x 28" D. 1990s. $650. *Courtesy of WFP.*

One of a pair of hand-carved rosewood tables with delicate brass inlay work and drawer in base. 19.5" H x 17" L. $300-500. *Courtesy of L'ambiance.*

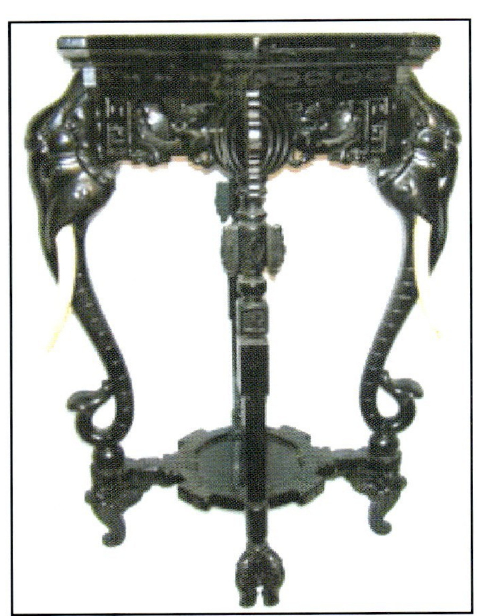

Carved (possibly mahogany) wood table with elephant head stanchions with ivory tusks. 36" H. c. 1890s. $500. *Courtesy of Arin Zablocki.*

Section 03.059 Greeting Cards

This subsection presents images of a few of what may be tens of thousands of greeting cards.

Comic paper greeting card. 6" H x 5.5" L. c. 1950s. $10. *From the author's collection.*

Reflective paper birthday card by Carlton Cards. 6" H x 6" L. 1990s. $3. *From the author's collection.*

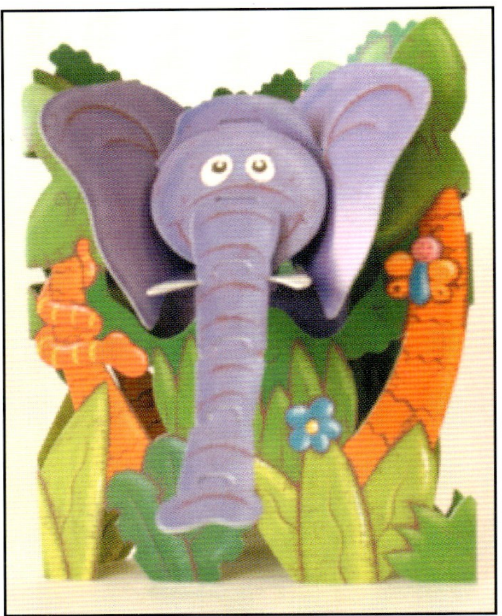

Swinging elephant 3-dimensional paper birthday card by Santoro Graphics Ltd of London, England. 7" H x 6" L x 7" D. 1998. $5. *From the author's collection.*

Sierra Club paper birthday card. 1990s. $4. *From the author's collection.*

Section 03.060 Hooks

This subsection presents images of elephant hooks, wherein the trunk is usually the functional "hook."

Brass belt hook. 4.5" H x 5" L. 1990s. $15. *Courtesy of Adele Verkamp*.

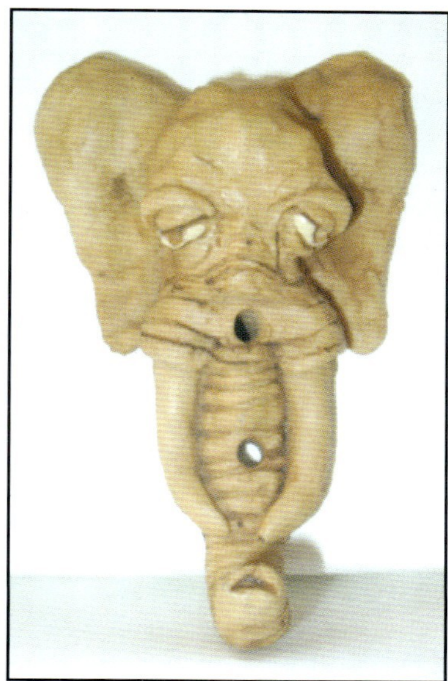

Painted wood finial on drapery rod by Atelier Co. 17" H x 11" L. 2000. $265 pair. *Courtesy of The Great Indoors*.

Resin hook, "Curious Critters." 6" H x 4" L. $10. *From the author's collection*.

Section 03.061 Hotpads

This subsection presents images of elephant hotpads.

Linen and polyester hot pad from WC Imports, Philippines. 7.25" H x 9" L. 1980. $5. *From the author's collection*.

Cotton Christmas motif hot pad by Ocean Desert Sales, Philippines. 9" H x 6.5" L. 1980s. $5. *From the author's collection*.

Section 03.062 Humidors and Tobacco Tins

This subsection presents images of humidors and tobacco tins sporting elephants.

Glass cigarette box by Greenburg Glass Works. 2.5" H x 3.5" L. 1920s. $50-75.

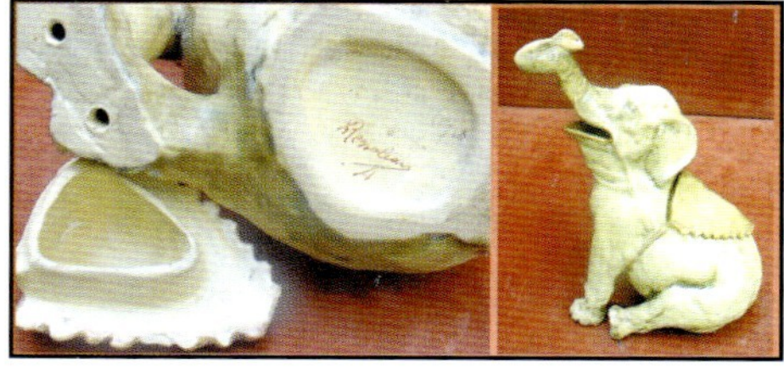

Ceramic humidor signed by Alfred Renoleau well-known in France for his Palissy ceramic works at the end of the 19th century. This piece seems to be an "humidor" with its triangular cap, but elephant's mouth is open, so it could be a figural pitcher too. 11" H x 7" L. c. 1880-1900. $450. *Courtesy of Denis Cordier.*

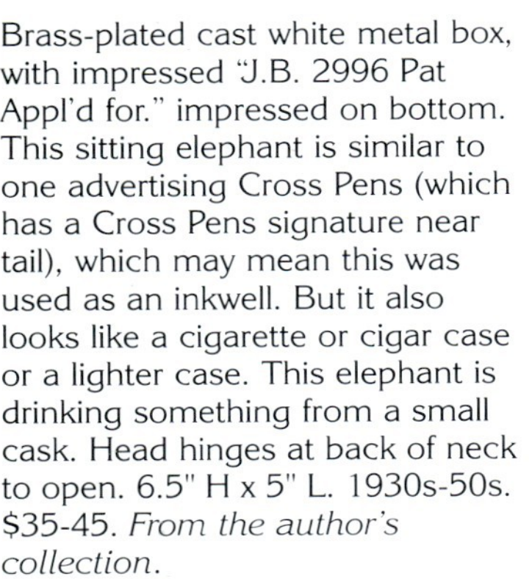

Brass-plated cast white metal box, with impressed "J.B. 2996 Pat Appl'd for." impressed on bottom. This sitting elephant is similar to one advertising Cross Pens (which has a Cross Pens signature near tail), which may mean this was used as an inkwell. But it also looks like a cigarette or cigar case or a lighter case. This elephant is drinking something from a small cask. Head hinges at back of neck to open. 6.5" H x 5" L. 1930s-50s. $35-45. *From the author's collection.*

Relatively rare ceramic humidor made by Abingdon Pottery, which closed in 1950. It has a full figure of an elephant, in black glaze, on the lid. The jar portion is fluted. The bottom is marked "606" and "Abingdon USA." 10" H x 6.25" L. c. 1940s-50. $500-1000. *Courtesy of Ed and Kate Meringolo, Sarasota, Florida.*

Majolica figural elephant humidor/tobacco jar. Impressed numbers on the bottom "6021" and "43." Dressed in a cutaway. This elephant tobacco jar was made in Germany or Austria. See *Horowitz, 154.* 7" H x 3.5" L. c. 1890s-1900. $300-400. *Courtesy of Danny.*

Rare Victorian ceramic tobacco jar from Austria marked "FGW" and numbered "148" on the bottom. Hand-painted lifelike depiction of decorated elephant with a basket for passengers on his back. His carter is a young Indian boy with a turban who is sitting on his head. See *Horowitz*, numbers 692 and 693. The mark is the same as #693 but the model is the same as #692, which has a different mark "CandH 389." Manufacturer is F. Gerbing who split from the original company in 1885. 7.7" H. c. 1890s. $300-400. *Courtesy of Alexander Zacke.*

Section 03.063 Incense Burners and Censers

This subsection presents images of incense burners. One prominent maker of incense burners featuring elephants that all collectors should know about is Vantines, a French firm that made several elephant burners. The Chinese also made many different burners with elephant motifs.

Signed brass French Vantines model 1265 incense burner. 6" H x 3" L. early 1900s. $60-85. *From the author's collection.*

Signed brass French Vantines model 1262 incense burner. 6" H x 3" L. early 1900s. $60-85. *From the author's collection.*

Rosemeade (also known as Wahpeton Pottery Company) incense burner that was also made in wine and rose; the company made pottery in North Dakota from 1940-1961. See *Sampson and Harris 1986, 106.* 4" H. $75-100. *Courtesy of Lorraine Miller.*

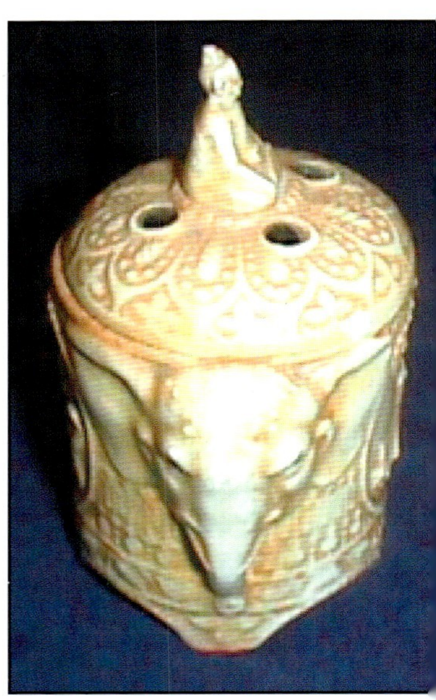

Oriental ceramic incense burner with Buddha sitting on top of the lid. It has a dancing lady on one side, a snake charmer with two cobra snakes on the other side, and elephant head handles. 4.5" pre-1950. $30. *Courtesy of Tim.*

Chinese bronze censer. 3" H x 3.75" L. c. 1890s. $40-50. *From the author's collection.*

Bronze Japanese incense burner. Enameled horizontal panels on front and back. Applied bronze finial appears to be a combination of an elephant and a scarab. Applied bronze handles resemble tigers with elephant's trunks. Provenance: Peter Ogden Collection. 3.75" H. c.1930. $300. *Courtesy of John M. Andersen, Jr.*

Japanese brass incense burner with pagoda. 5" H. c. 1950. $15-20. *From the author's collection.*

Tibetan or Chinese bronze incense burner, ornately caparisoned. 8" H x 8.25" L. Late 1800s. $200-300. *From the author's collection.*

Signed brass French Vantines model 1205 incense burner with its original tin insert. 5.5" H x 6" L. early 1900s. $125. *Courtesy of Darrell.*

Section 03.064 Inkwells

This subsection presents images of inkwells with an elephant motif. Inkwells are individual containers for the ink itself, often positioned within a fancy stand or receptacle with one or more accessories and decorative accouterments (like elephants!). Franklin and Jean Hunting (*Inkwells, 2000*) say that most U.S. inkwells were made from the mid-1800s through 1930s. Since ink has been used from about 2500 B.C., the first "inkwells" are ancient, and were probably made of bone, horn, or mineral.

Prior to 1600s few could read or write, so not many inkwells date to that period. Those that do were used mostly in monasteries, government offices, and in academic institutions. By the 1700s, the Industrial revolution sparked a reading and writing trend among most people, so inkwells were needed and therefore their manufacture increased. Inkwell motifs followed the general arts motifs during the succeeding decades: Arts and Crafts 1861-1920, Art Nouveau 1890-1910, Art Deco 1910 to about the 1930s. The demise of inkwells started in 1884 when Waterman introduced the fountain pen, which contained its own ink. Most inkwells were gone by 1950 when the Post Office in the US replaced theirs with ballpoint pens.

Silver-plated elephant and rider inkwell. The rider tips back to reveal the glass ink well. Probably Asian. 4.75" H x 3.25" L. c. 1920. $100-150. *Courtesy of Derek Holloway.*

Victorian inkwell, bronze elephant on metal stand painted in a marblesque finish. c. 1900. $150-200. *Courtesy of Jack and Sharon Timmer.*

Bronze lying elephant inkwell with turbaned rider; rider's hand is chained to the elephant's ear. 6" H x 5" L. c. 1900. $100-125. *From the author's collection.*

Unmarked, matte glass inkwell; probably French. 4" H x 4" L. c. 1900-1910. $200-250. *Courtesy of Andrew Mattijssen.*

Wooden inkwell. $100-150.

This spelter with gold wash inkwell is an elephant head with a glass insert and a monkey finial. It is marked "J.B. 356" and was made in America. 4" H x 6" L. c. 1870. $225-275. *From the author's collection.*

Brass or bronze inkwell with name inscribed on the front of this piece that appears to be "Jaremere" or "Faremere." 3" H x 5" L x 3.25" D. $35-50. *From the author's collection.*

A cast iron inkwell with a raised display of elephants in the jungle on the base. This is a double inkwell, with the original glass inserts and sliding metal covers. The small elephant statue in the middle of the inkwell is non-magnetic. 3.5" H to elephant's back. 8.25" from well to well, 9.5" in the front where the pens rest. c. 1920s-1930s. $150-200. *Courtesy of Jeff and Joann Drown.*

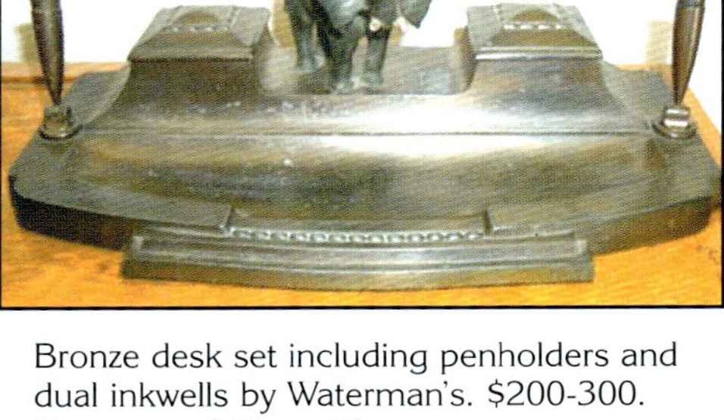

Bronze desk set including penholders and dual inkwells by Waterman's. $200-300. *Courtesy of Diane Moore.*

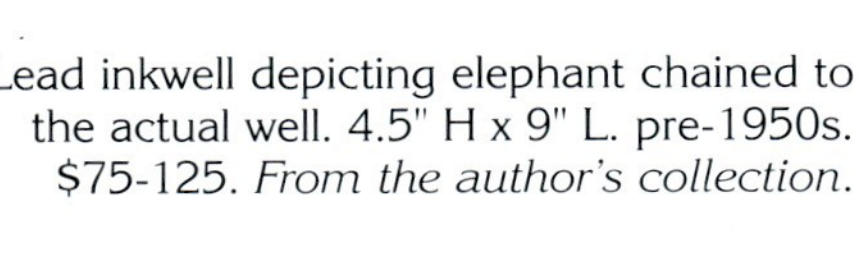

Lead inkwell depicting elephant chained to the actual well. 4.5" H x 9" L. pre-1950s. $75-125. *From the author's collection.*

Section 03.065 Ivory Elephants

This subsection presents images of elephants made of ivory. Most ivory elephants are made from carved elephant tusks; but there has been a surge in elephants made from hippo and mammoth "ivory" recently. Small carvings with intricate etchings, with a hole in the figure, are known as Netsukes; they are in their own category. Staining ivory with different colored teas was done to accent detailed carvings and is used to color the dark side of a chess set.

This category may be controversial to some of you. Since most ivory carvings resulted from the death of an elephant, I hesitated and pondered including this type of elephant collectible. Yet ivory is a material, and many ivory elephant things have been imported into the U.S. before a ban on ivory importation was in place. And I feel that bona fide elephant collectibles, even those made of ivory, have a place in any "encyclopedia" of elephant collectibles. Another way to look at it is that virtually all materials are obtained by some sort of damage to our surroundings; we take minerals from the land, we cut down trees, and we make clothes, purses, etc. out of animal hides and fur. I personally do not wish to see one more elephant, or any animal for that matter, sacrificed merely for the benefit of creating some luxury bauble.

The classic ivory "bridge" of elephants. 18" L. $300-500. *Courtesy of Henry J. Hebing.* (E2D4)

"Mountain of Elephant," four tiers of ivory elephants on a wood platform. 2.5" H x 2" L. 1970s. $45. *Courtesy of Ted Capell.*

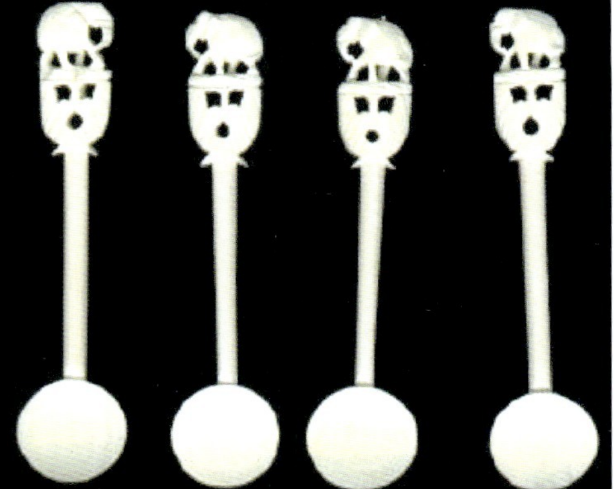

Set of four ivory salt spoons. $50-75. *Courtesy of Linda Ferreira.*

Finely detailed Japanese Okimono elephant. Fogen Boatsu seated on an elephant carrying a small child with three karako climbing up to join her. The trappings of the elephant are inlaid in shibayama style with mother of pearl, hardstones, and stained ivory. It was carved in the Meiji period from one piece of ivory; there are no joints or additions. It is signed in cursive underneath on a large oval mother-of-pearl inlay. 6.5" H x 4.5" L. 1880s. $250-300. *Courtesy of Glenn.*

Large, detailed carved ivory elephant on carved rosewood stand. 9" L x 5" H. c.1950s. $800-1000. *Courtesy of Midwest Estate Buyers.*

Ivory elephant with age patina. $100. *Courtesy of Henry J. Hebing.*

Detailed scrimshaw of elephants on a hippo tusk. 1990s. $500. *Courtesy of David Boone at Boone Trading Company, Inc.*

Detailed ivory elephant. The ivory has been stained with a tea dye and is inlaid with pieces of coral and turquoise. The wood stand is inlaid with silver accents. From China. With stand: 4.5" H x 3.5" L x 2" Deep. c.1970s. $200-300. *From the author's collection.*

Section 03.066 Jewelry

This subsection presents images of jewelry including bracelets, rings, cufflinks, necklaces, watches/watch fobs, and pins or brooches. Some may consider belt buckles and Netsukes jewelry too, but those things were put into their own sections. Some of the more sought after elephant jewelry includes items made by: Cartier (whose gold line of elephant jewelry ranges from $4000-$40,000!), Ciner, Hattie Carnegie, and Alexander Korda.

Bronze pendant from an elephant blanket used during the Mogul Empire in India features a ceremonial elephant in relief. 1.75" H x 1.75" L. 1700s. $20-30. *From the author's collection.*

Stainless steel necklace with crystal eye. 3.5" H x 4" L. 1970s. $15. *Courtesy of Adele Verkamp.*

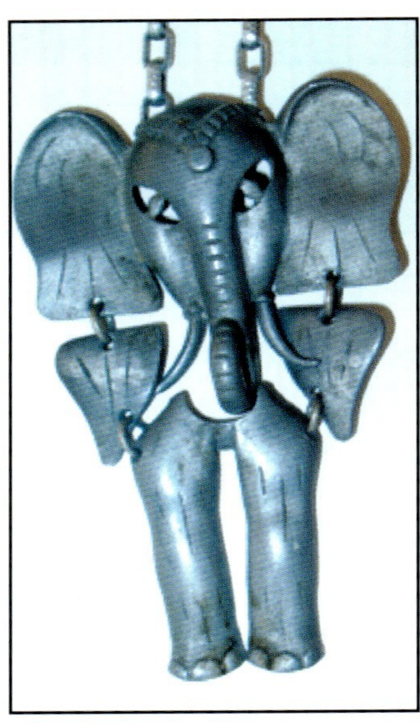
Pewter articulated necklace. 4" H x 2.5" L. 1960s. $15. *From the author's collection.*

Rare pin by Alexander Korda. Raised design of elephant in "spoon" portion of pin with plastic, possibly Bakelite, horns in top portion. 3" H x 1.5" L. $150-200. *Courtesy of KD's Collectibles, Lakeland, Florida.*

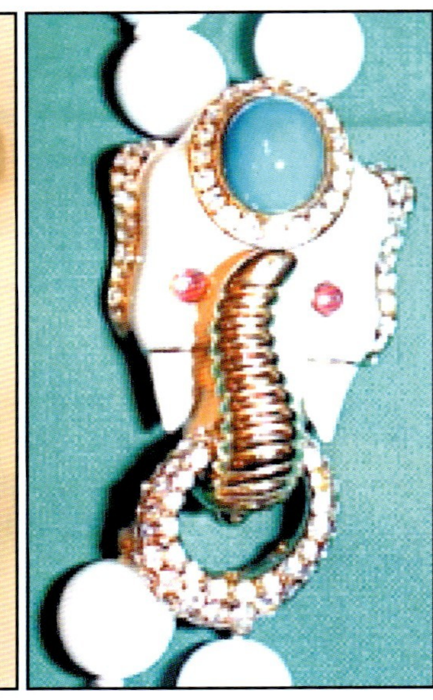
Elephant head enameled clasp for a choker/necklace, accented by clear crystals, a turquoise-colored headpiece, and ruby-colored eyes. The back of the elephant head bears a plaque with "CINER" and a copyright symbol. Ciner began making high quality costume jewelry in 1931. 1.5" H x 1" L. 1940s. $100-150. *Courtesy of Karen Beltz.*

This coral elephant on lapis sphere is decorated with 18 karat gold and tiny diamonds and jade stone. A bale has been attached so that the piece could be worn as a pendant. On the back is the original pin. Including the bale the piece is 2.5" H x 1.5" L. $250-400. *Courtesy of B-Street Motion, in San Rafael, California.*

Orange earrings made from Bakelite. 1940s. $25. *Courtesy of Carol Hill.*

This Hattie Carnegie Co. quartz brooch is caparisoned in silver with channel-set rhinestones. The brooch is signed "Hattie Carnegie" in a disk on the back. Hattie Carnegie made several other elephant pins dating to the 1920s or so. 1.375" H x 2" L. 1960s. $60-90. *Courtesy of Cathy Sardella.*

Pin signed by Ciner. The elephant is gold-plated base metal, with rhinestones, accented with black enameling,. The eye of the elephant is a faceted green stone and the tusks are enameled to look like ivory. 1.25" H x 1.85" L. $100-150.

Old elephant watch fob with "Lucky Elephant Elphinstone of Baltimore Md." 1.75" H x 1.25" L. $25.

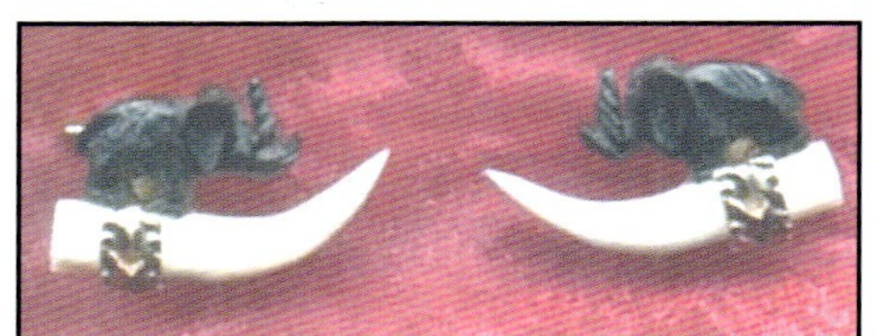
Carved ivory cufflinks with elephants attached made by Swank. 1940s-50s. $50-75. *Courtesy of Debra A. Malarkey.*

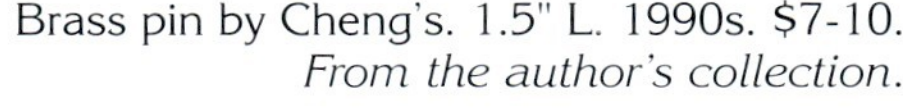
Brass pin by Cheng's. 1.5" L. 1990s. $7-10. *From the author's collection.*

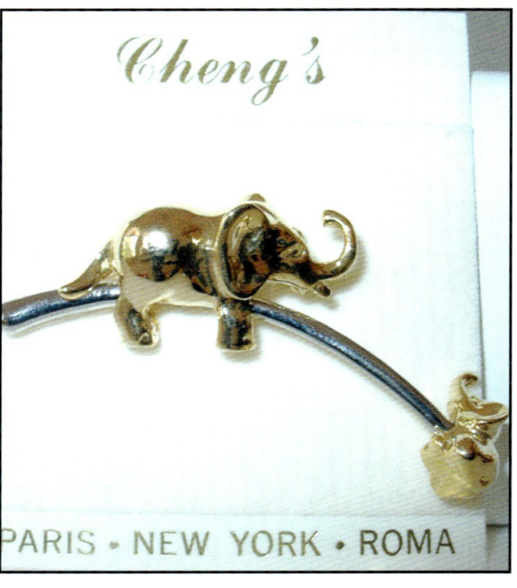

Section 03.067 Key Rings

This subsection presents images of elephant key rings.

Enameled brass key ring with "SWIB Taiwan" on back. 1.75" H x 2" L. $3. *From the author's collection.*

Plastic key ring with built-in ruler. 1.625" H x 2.625" L. $3. *From the author's collection.*

Section 03.068 Knives and Cutlery

This subsection presents images of knives that have elephant motifs. Arthur Court is one maker of elephant knives (and other kitchen dishes/accessories etc.). Case started putting elephant etchings onto their blades in 1975; they have become sought-after collector's items. And the Franklin Mint produced a great Serengeti African Elephant Knife in limited edition elephant in 1998 valued at about $200.

Brass and steel Tibetan Dergu ritual knife has t-handle with elephant head guard and crown pommel. 7" H x 7" L. 1990s. $20. *From the author's collection.*

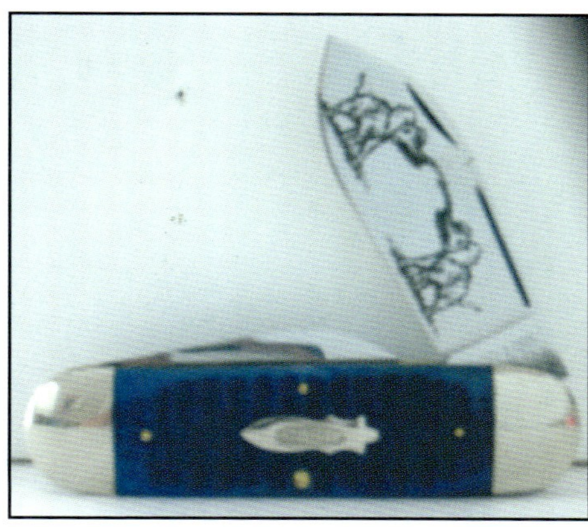

W. R. Case and Sons (Bradford, Pennsylvania) made this "Classic Sunfish" knife in a limited edition with double elephant etching in blade and blue bone scale pattern in handle. Pattern #62050 SS; one of only 9 produced. Case started the elephant etchings in 1975. 8" H (open) x 1" L. 1996. $150-250. *From the author's collection.*

Arthur Court aluminum bread knife. 12" L. 1990s. $25-30.

Section 03.069 Labels

This subsection presents images of elephants on labels. Of course, these are freestanding labels; many other things in this book, like cans or tins, have a label attached. Other labels I know of include Elephant Margarine, Jumbo Grapes, and Elephant Brand Corn.

Primeros inner cigar box label with Dept. No 16712. 7" H x 10.5" L. $7-10. *From the author's collection.*

Hotz Hotels, India, baggage label. 4" H x 5" L. $15. *From the author's collection.*

Rambagh Palace, Jaipur, India, luggage label. 4.75" D. $20-25. *From the author's collection.*

Strength Santa Paula Orange Assoc. Sunkist crate label. 10" H x 11" L. 1936. $30-40. *From the author's collection.*

Elephant Brand firecracker package label from Macau, China. 5" H x 4" L. $10. *From the author's collection.*

Cigar label made by F.H. Berning Co. for Ivory King cigars. c. 1900. $200-300. *From the author's collection.*

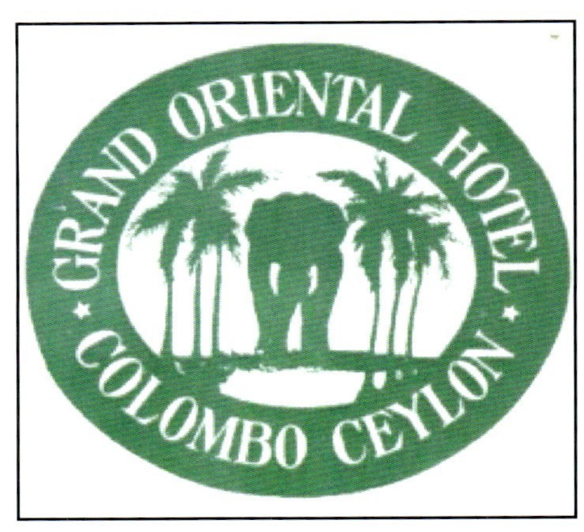

Luggage label from the "Grand Oriental Hotel" in Columbo, Ceylon (Sri Lanka). 4.375" H x 5" L. c.1930s. $25. *From the author's collection.*

Round luggage label from the West African Airways Corporation (WAAC), established in 1946 as the national airline of British West Africa. The airline's logo was a winged-elephant and their motto became "Where Elephants Fly." Existed until 1958 (name still exists in Nigeria). 4.5" D. 1950s. $50. *Courtesy of Ben Guttery.*

Section 03.070
Lamps (Oil and Electric)

This subsection presents images of lamps in the shape of an elephant. I show both oil and electric lamps. Many of the most interesting elephant lamps were produced either in the Art Deco period or shortly thereafter. In addition, several of the major bronze art firms produced lamps out of their elephant wares, including Frankart.

Glass oil lamp. 5" H x 5.5" L. 1990s. $30-35. *Courtesy of Adele Verkamp.*

Ornate bronze oil lamp with removable bowl that fits into urn on top of elephant's back. Elephant is in parade regalia and urn has lion masks and caryatids, and receptacles for lighting wands. 6.5" H x 7" L. 1800s. $150-200. *From the author's collection.*

Miniature oil lamp with porcelain elephant with a small dog on a platform adorned with green leaves. The howdah is trimmed in pink and there is an orange tassel on either side. White Bristol shade and Sparr Brenner burner. 9.75" H (with shade). 1890-1910. $750. *Courtesy of Marjorie Hulsebus.*

Antique metal lamp. 15" H x 17" L (with shade). c. 1930s. $75-100. *From the author's collection.*

Satsuma lamp with white glass globe. 11.5" H x 5" L. $40-50. *Courtesy of Adele Verkamp.*

Art Deco lamp is spelter with a bronze finish and depicts a family of elephants. Ruby glass globe has flames all around in relief. There is an on/off switch in the back. 8" H x 3.5" L; 13.5" H to the top of the shade. c.1930s. $75-100. *Courtesy of Linda Yakiwchuk.*

Cast lamp and ashtray combo with original opal milk glass ash receiver by Frankart Inc., of New York. Original came with parchment shade. 1928. $125-200. *Courtesy of Collection of Jeff Leegood, Decollectibles.*

Terra cotta Roman oil lamp. 1" H x 3.25" L. est. 300 AD. $150-250. *Courtesy of Ardre, of Arctic Coin, Ottawa.*

Section 03.071 Light Bulbs

This subsection presents images of elephants as light bulbs. There is another light bulb with a GOP elephant filament in the Political/GOP section.

Elephant night lamp in orange from FantasTick USA Inc. Hand-painted, 110v/UL Approved, 6' Cord w/Insulated Switch and Polarized Plug, Std. Night Light Bulb (Replaceable) #8030015. FantasTick has other elephant bulbs in their line. 7" H. 1990s. $10. *Courtesy of FantasTick USA Inc.*

Lightbulb with a metal elephant filament inside. 40 watt. c.1930s-40s. $65.

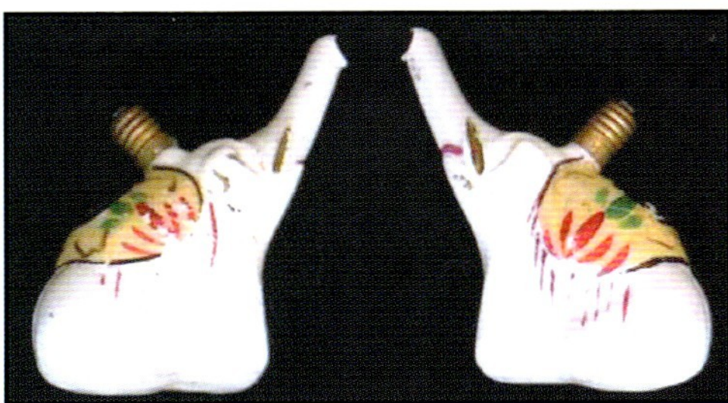

Elephant-shaped Christmas bulb marked "JAPAN." 4" H x 1" L. c. 1950s. $15. *Courtesy of Pat.*

Wiener Workstatte Hagenauer brass "match striker" lighter. An Art Deco elephant balances the match holder and striker on the tip of his trunk. Marked: "Made in Austria" and has the circled WHW symbol. 2.75" H x 2.5" L. c. 1920s. $150. *Courtesy of Peggy Litchfield.*

Section 03.072 Lighters and Match Strikers

This subsection presents images of lighters. In *Cigarette Lighters*, by Stuart Schneider and George Fischler, it is noted that the first lighters were made about when cigarettes were popular in the mid-1850s. Several elephant lighter designs are from the Art Deco period (1920s). There are many striker lighters with an elephant motif. A variety of liquid/gas-fueled lighters were also made until smoking became less popular in the 1970s due to increasing health concerns. The earlier elephant lighters likely will become more desirable as they become rarer.

The two most popular lighter manufacturers are Zippo and Ronson. Both have produced elephant lighters at very different times and in different styles. Ronson's great elephant designs were made in the first half of the 1900s, and reflected the need for striker lighters. Zippos are more recent and usually involve an elephant etching on the side of one of their fluid lighters.

Silvered cast elephant head lighter. 3.5" H x 4.5" L. $25-35. *From the author's collection.*

Signed Vienna bronze elephant match striker and matchsafe marked "AUSTRIA - GESCHUTZT." 4.75" H x 3" L. $300-450. (E2D4).

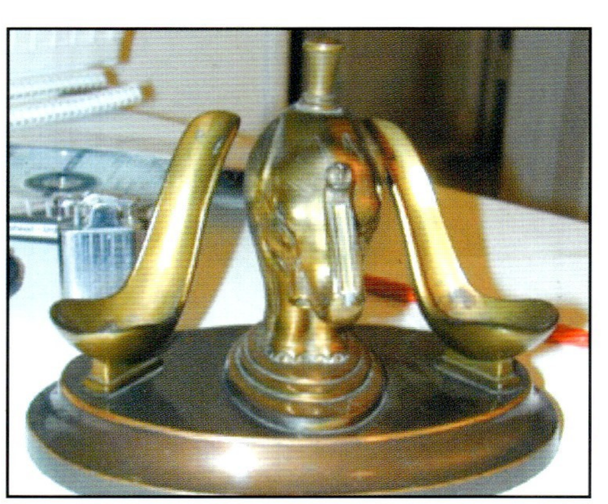

Rare Ronson Striker lighter and 2 brass pipe holders on base. 6" H x 7" L. early 1920s. $1500. *Courtesy of Scott Moore.* (E2D4).

Bronze Ronson striker lighter as walking elephant; wand is in the elephant's back. c.1910-1920s. $800. *Courtesy of Scott Moore.* (E2D4).

Bronze table lighter with lift-arm. 3.25" H. 1930s. $30-40. *From the author's collection.*

Art Deco chrome-plated striker lighter made by the Ronson Art Metal Works Inc. Original tusks and a felt bottom with a "Ronson" label. A bronze colored version of this lighter is worth up to $600. 5.375" H x 3.75" L. 1935. $125-175. *From the author's collection.*

Rosenfeld by Florenza lighter. 2" H. 1940s-50s. $25-35. *Courtesy of Denise Massaro.*

Ronson lighter with sterling silver engraved elephant image on a black background. The other side has a grand palace below a mountaintop. Under the palace is some writing and on the elephant side are the letters "g p" in script. The bottom of the lighter says," RONSON - MADE IN ENGLAND." 1.875" H x 1.5" L. $40-60.

Die-cast metal and plastic gas and electric (battery-powered ignition) table lighter by Rony of Japan. 10" H x 6" L. $25-35. *From the author's collection.*

Section 03.073 Liquor Bottles

This subsection presents images of elephant bottles used to store liquor. The most common liquor bottles are probably those produced under the Jim Beam and Erza Brooks names. They commemorate GOP-related occasions and other events. One of the most sought-after Beam bottles is the Spiro T. Agnew elephant of 1970 valued at over $500! Oliphant also made various elephant liquor bottles including one for crème de cacao and one for their vodka.

Bertha liquor bottle. Bertha was at the time the world's most talented performing elephant. She performed nightly for theater restaurant patrons at John Ascuaga's Nugget Casino in Nevada starting in 1958. This decanter honored Bertha and was made by Erza Brooks to hold 12-year bourbon. 10.5" H x 11" L. 1970. $30-40. *From the author's collection.*

Parfait amour art glass liquor bottle from Bols of Holland with cork stopper and blue liquor. 2.5" H x 4" L. 1960s-70s. $60-100. *Courtesy of Yolaine Gagnè, Quèbec, Canada.*

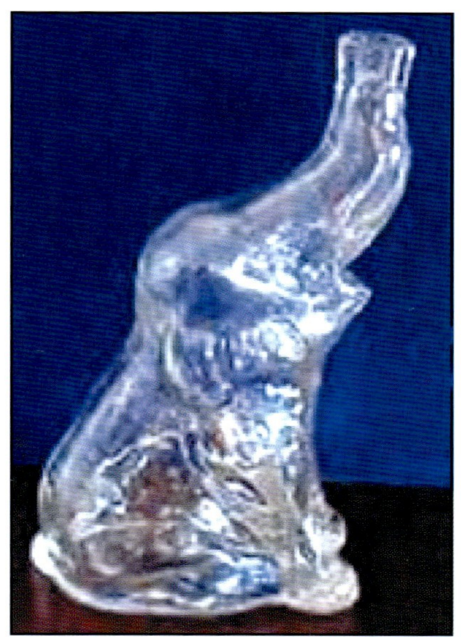

Elephant clear glass whiskey bottle by Fenton, considered rare. The printing on the bottle reads "Federal Law Forbids Sale or Re-Use of this bottle" and "R 344-95-5." 8" H. 1935. $250-400.

Ski Country Straight Bourbon whiskey decanter, Barnum Festival Est. 1949 Bridgeport Connecticut, Designed for Delibro Specialties, Foss Company Golden Co. Japan. 11.5" H x 4" L. 1973. $35-45. *From the author's collection.*

Jim Beam liquor bottle elephant is a boxer. 11.5" H x 6" L. 1964. $30-50. There are several other Beam elephant whiskey bottles including: Football with Player in 1972, Clown in 1968, Republican in 1976. Most range from $10-50, with notable exceptions including the 1970 Spiro Agnew ($1000) and 1972 Miami Beach bottles ($400-600). *Courtesy of Adele Verkamp.*

Hull Pottery pink and white elephant liquor bottle with original cork. Incised "USA" on bottom. It was made for Leeds liquors; some still have the original state liquor labels on them, which can increase the value by $5-15. 7.5" H x 5" L. c.1940s. $75-110 in *Roberts, 1997. From the author's collection.*

Apricot Liqueur bottle by Drioli. 4.5" H x 3" L. 1950s. $20. *Courtesy of Cathy and Rick Gross.*

Section 03.074 Magazines and Covers

This subsection presents images of magazine covers featuring one or more elephants.

National Geographic "Elephant Talk" magazine cover. August 1989. $5. *From the author's collection.*

Life magazine elephant drawing called "Four In Hand" by Harry B. Neilson. 28 October 1897. $20. *From the author's collection.*

Wargamers Digest magazine cover with an article on the use of the elephant in ancient warfare. Cover picture is a Hinchcliffe model painted by Dick Zimmerman photographed by William Kojis. 11" H x 8.5" L. April 1975. $7-10. *From the author's collection.*

Field and Stream magazine cover. 11" H x 8.5" L. January 1951. $10. *From the author's collection.*

Nature Magazine cover. 11" H x 8.5" L. June 1932. $8-10. *From the author's collection.*

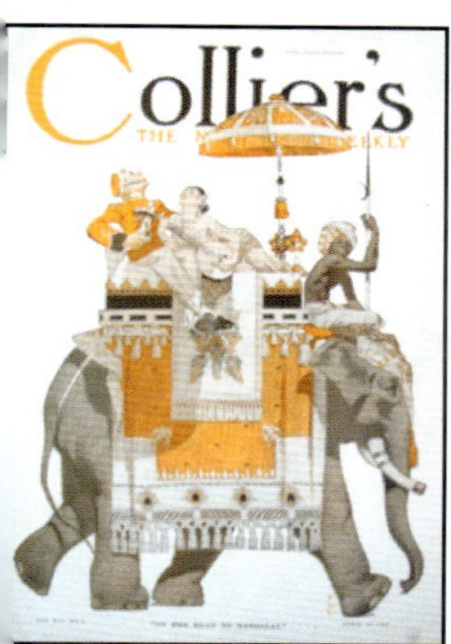

Collier's magazine cover; drawing called: "On the Road to Mandalay" by F.X. Leyendecker. 15" H x 10.5" L. 18 April 1908. $10. *From the author's collection.*

Saturday Evening Post magazine cover; elephant drawing by F.X. Leyendecker. 17, October 1936. $10. *From the author's collection.*

Harper's Weekly magazine cover with elephant drawing by T.S. Nast. 15" H x 10" L. 20, November 1880. $35-50. *Courtesy of Adele Verkamp.*

Computer Design magazine cover. 11" H x 8.5" L. March 1995. $5. *From the author's collection.*

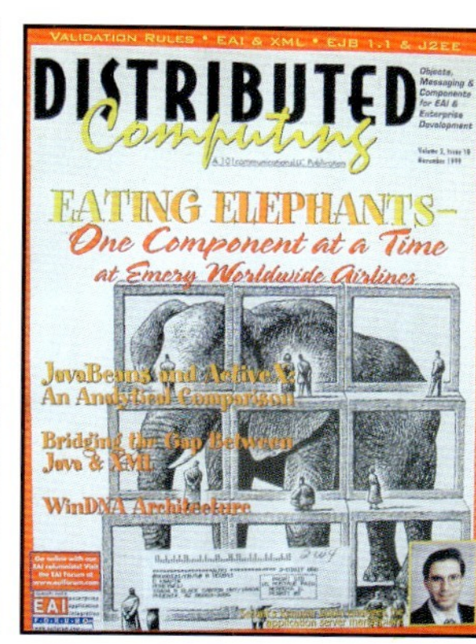

Distributed Computing magazine cover. 11" H x 8.5" L. November 1999. $5. *From the author's collection.*

Section 03.075 Matchbooks and Matchbox Labels

This subsection presents images of matchbook covers and matchbox labels with elephant images on them.

A trademarked Jumbo on a box for safety matches from Sweden. 1.25" H x 2.25" L. $3. *From the author's collection.*

Sorunku matchbox carton cover. 2.25" H x 3.5" L. $3. *From the author's collection.*

L'Elephant Blanc Salon De Barbier, Canada matchbook cover, Eddy Match Co. $2. *From the author's collection.*

Windsor Bourbon matchbook cover. $2. *From the author's collection.*

Section 03.076 Matchsafes and Match Holders

This subsection presents images of matchsafes and match holders. Matchsafes were used in Victorian times when wooden, non-safety matches were the only light in town. Storing them in a metal container small enough to carry around, ensured they would not ignite at inopportune moments (in your pocket!). As with other items, progress eliminated the need for these artifacts and they are now quite valuable.

Nickel-plated brass matchsafe with glass eyes and bone or ivory tusks. Made in Europe. 1.2" H x 2" L. c. 1885-1900. $425-475. *Courtesy of Ron Burke.* (E2D4).

Sterling silver matchsafe with garnet cabochon eyes and a ribbed area under his chin for use as a striker. This is a reproduction of a Victorian brass matchsafe of the same design shown previously. 1" H x 2" L. 1990s. $50-75.

Bronze elephant atop a matchbox holder. Marked on its belly: "K and C Co." Plaque on box below elephant reads "FORT KNOX, KY." Inside the piece was an old wooden matchbox produced by Tartan Safety Matches. Distributed by Alfred Lowry and Bro., Philadelphia Pennsylvania. $20-30. *Courtesy of p.z. gluck.*

Commemorative Indian brass and copper matchbox holder made for Rudyard Kipling's death; has Kipling's quote: "A women is only a woman, but a good cigar is a smoke," Kipling's name and DOB-DOD inscribed (1865-1936). 1.6" H x 2.25" L. 1936. $25-30. *From the author's collection.*

Brass and pewter matchbox holder. 2.75" H x 1" L. c. 1940s. $20. *From the author's collection.*

Silver-plated metal matchsafe with ivory or Bakelite tusks. Striker is on his back with hinged neck. 1.5" H x 2.25" L. c. 1910. $300-450. *Courtesy of Sharyn Proctor of Willow Antiques, South Island, New Zealand.*

Section 03.077 Mirrors

This subsection presents mirrors with an elephant image.

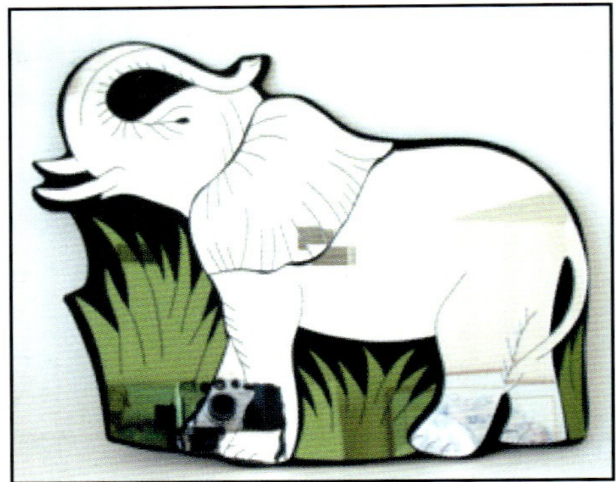

JJ Blinkers "die-cut" mirror. 12" H x 17" L. 1970s. $20. *From the author's collection.*

Painted and engraved mirror. 11" H x 14" L, w/o frame. 1980s. $10. *From the author's collection.*

Section 03.078 Miscellaneous

This subsection presents images of a variety of elephant items that do not readily fit into the other subcategories. They may be made from an unusual material, covered with something unusual, or just an uncommon item.

Carved Indiana limestone elephant head downspout placed on sculptor/carver Walter S. Arnold's home. 24" H x 32" L; weighs 500 pounds. 2000. Quote on request. *Courtesy of Walter S. Arnold.*

Nicely detailed rubber-like magnet shown in the Collection-O-Magnets image later in this category. 2.75" H x 3.25" L. $3. *From the author's collection.*

Placemat of rubberized material. 12" H x 18" L. $3. *From the author's collection.*

Collection-O-Magnets including (from top, right-to-left): Living Stone 1987, clay, ceramic, rubber GOP, Clay Critters, plastic card 1982 with words: "Don't Forget," plastic with ruler around waste from Russ Berrie with words: "You Stuff You Puff." brass, plastic mom and baby, ceramic, Proudline #3984 Lucite card with words: "I never met a carbohydrate I didn't like," pewter, ceramic. All 3" x 3" or less. 1970s-1990s. $2-10. *From the author's collection.*

Stained glass window hanging. 5" H x 3.5" L. 1990s. $5-10. *From the author's collection.*

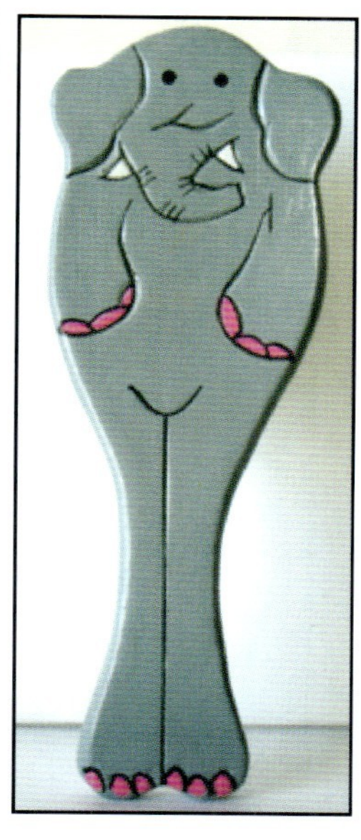

Wood hairbrush. 7.5" H x 3" L. $5. *Courtesy of Adele Verkamp.*

Clay toothbrush holder by Nancy Ecrlono. 6.5" H x 4" L. 1998. $10-15. *Courtesy of Adele Verkamp.*

Cardboard mask on wood handle. Words on back read: "BeachCombers International 1981 Taiwan." 12" H x 8.5" L. 1981. $5. *From the author's collection.*

Hotbots Woof and Poof "Elle Phunt" hot water bottle. 16" H x 11" L. $15-20. *From the author's collection.*

Wind Catcher by Colore's Intl. USA. 23" H x 21" L. $7 *From the author's collection.*

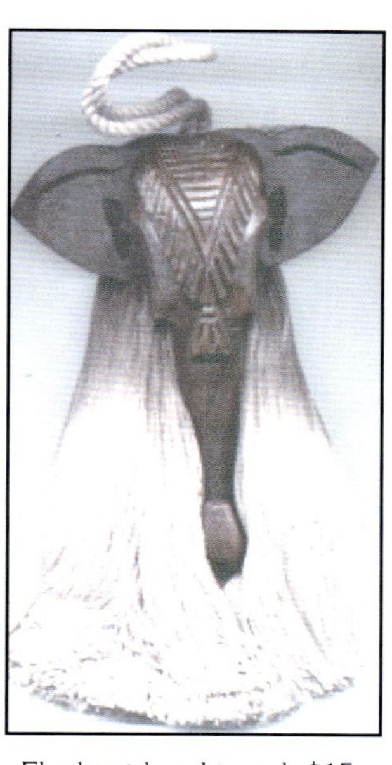

Elephant head tassel. $15. *From the author's collection.*

Plastic Halloween sucker display. $5.

Clothes hanger; fuzzy cloth over wire. 5" H x 12" L. $5. *From the author's collection.*

Horse brass used to decorate a horse. Depicts one of the famous pair of elephants kept at the London Zoo at one time: Jumbo and Alice. Horse brasses began as pure decoration hundreds of years ago, and the idea carried over to the British Empire as it spread to the Indian sub-continent. Larger brasses that were worn by the great Shires were also worn by the Indian elephants.1930s-50s. $25-50. *Courtesy of Paul L.*

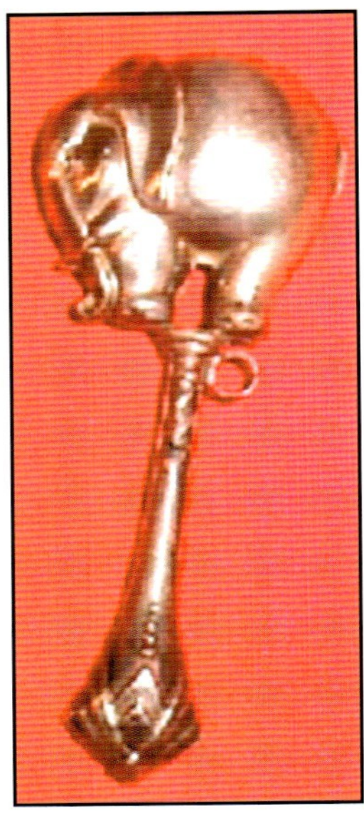

Brass mask inlaid with coral and rhinestones. 1940s-50s. $50-100. *Courtesy of Recollections, Ltd. of Elk Grove Village, IL.*

Finely articulated sterling silver baby rattle engraved "925" and "Thomas." 6" H x 2" L. 1890-1910. $50-75. *Courtesy of Wanda J. Crews, Ph.D.*

Large brass elephant telephone with touch-tone keypad. 14" H x 13" L. c.1970. $50.

Gems of the Earth resin water globe model @7017 by Westland; made in China. $30-50. *Courtesy of Lenny (Lenora) Salandi.*

Luckyphant hand-painted slot machine with playing elephants made by Banberry Designs, Inc., which makes many elephants. $15-25. *Courtesy of Banberry Designs, Inc.*

Metal reproduction of Mughal-period foot scrubber "jahma" has embedded loose metal balls that jingle to warn others that a room is occupied. 2" H x 2" L. 1990s. $10. *From the author's collection.*

Animal-embellished bath towel set. Knight Ltd. 2000. $18. *Courtesy of The Great Indoors.*

Section 03.079 Molds and Cake Pans

This subsection presents images of molds and pans used to form the shape of an elephant with a food, like chocolate or cake.

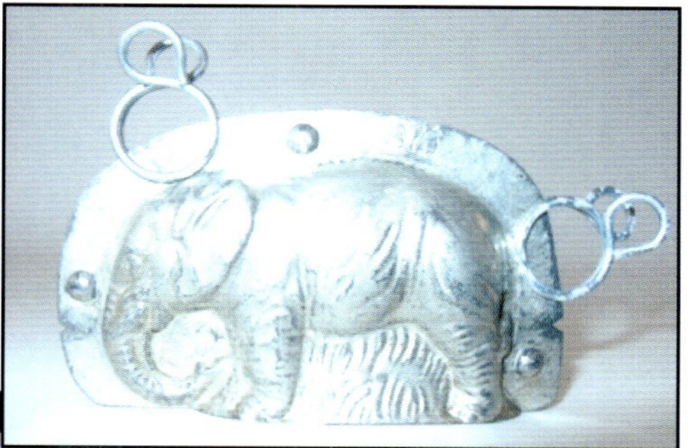

Belgian chocolate mold made by Anton Reiche. 2.5" H x 4" L. c. 1930. $60-75. *From the author's collection.*

French Sommet chocolate mold. $30.

Chocolate mold. Marked with an "E" trademark and numbered "5088." Made in USA. 7" H x 9" L x 4" D. $30.

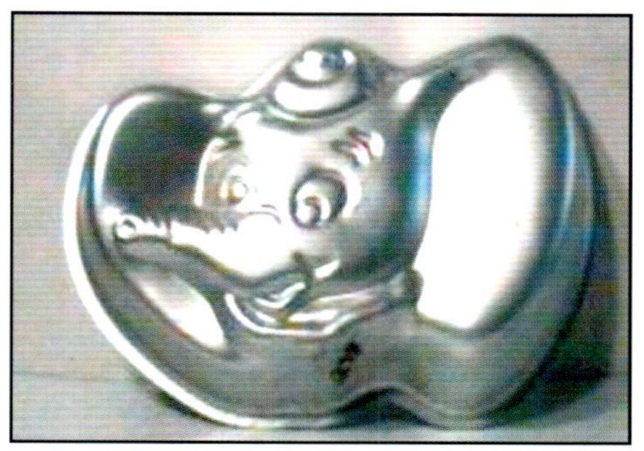

Aluminum cake pan of Dumbo. $10-20.

Pewter mold made by Cadot Compiegne, a French ice cream mold manufacturer. Victorian era. $100-150+.

Ironstone elephant mold, made in England. 4.25" H x 6.75" L. c. 1900. $200-300. *Courtesy of Diane Gellen.*

A two-sided chocolate mold of an Indian elephant. It is marked "16160" and is made by the Dutch mold makers JKV Tilburg. 4" H x 6.375" L. $50-75.

Section 03.080 Movies and TV Shows

This subsection presents images of movies and documentaries. Another one, not included here, is PBS's Special called "Living Edens" which is about the elephants living in the Anamalai Range including Elephant Mountain in India.

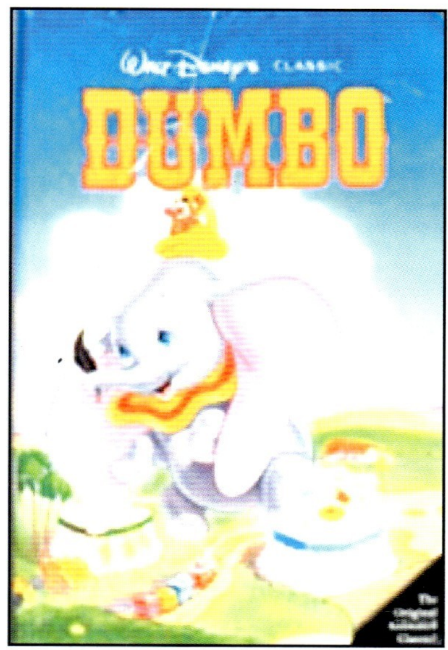

VHS Cassette of Disney's 1941 film "Dumbo." $10. *Courtesy of Nancy.*

VHS Cassette of Nova's 1990 elephant documentary: "Can the Elephant Be Saved." $10. *From the author's collection.*

Section 03.081 Music Boxes

This subsection presents images of elephant music boxes.

Porcelain music box made in Japan. 4.5" H x 5.25" L. $20. *From the author's collection.*

Napcoware music box/planter, plays 1817 Schubert's Lullaby. 4.5" H x 6.5" L. $10. *From the author's collection.*

Eden stuffed-elephant music box with moving head. 8" H x 9" L. c.1940-1950s. $10-15. *From the author's collection.*

Schmid (went out of business in 1995) tuba-playing elephant music box made in Japan. $15-20. *Courtesy of Carolyn Weber.*

Porcelain music box by Flambro made in Taiwan; plays "Talk to the Animals." 6" H x 3.5" L. $10. *From the author's collection.*

Section 03.082 Napkin Holders and Rings

This subsection presents images of napkin holders and napkin rings. Some of the most sought after elephant napkin rings are made of Bakelite.

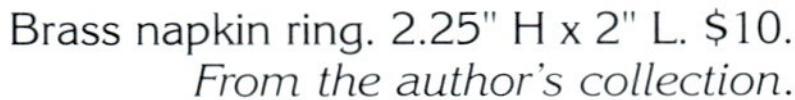

Brass napkin ring. 2.25" H x 2" L. $10. *From the author's collection.*

Hand-painted wooden napkin ring by Victor Bonomo Inc., New York, New York. Model #8004. 3.75" H x 2.5" L. $3-5. *From the author's collection.*

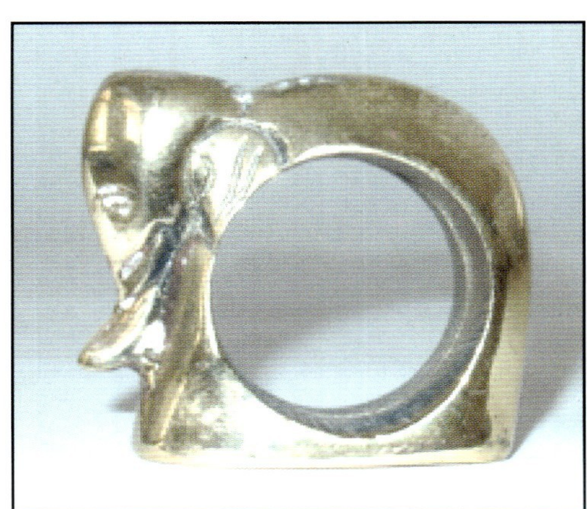

Section 03.083 Netsukes

This subsection presents images of small figurines with a hole through them called Netsukes. These small, often very detailed sculptures, were originally used in Japan as a button-like fixture on a man's sash, from which other belongings were hung. They are now often placed on a chain and worn as jewelry. The same expression of hesitation I made in the subsection on Ivory Elephants applies here too, since many Netsukes are made of elephant ivory. However, there are also many Netsukes made out of different woods, metal, ceramic, and horn and bone of other animals like mammoth (already extinct) ivory , and hippo bone.

Mammoth ivory netsuke elephant standing on a large ball and balancing another small ball on its trunk. Has onyx-inlaid eyes and is signed by the artist on a mother-of-pearl inlay on the bottom of the ball. The cord holes are in the elephant's hip. 2.75" H x 1" L x 1" D. 1990s. $75-125. *Courtesy of Aguanga Sunshine Imports.*

Polychromed ivory netsuke by artist Meike. 1.75" H x 2" L. 1970. $2900. *Courtesy of Michael Levin of Peking Art Importers, New York.*

Mammoth ivory netsuke elephant family. 1990s. $75-125.

Section 03.085 Nutcrackers

This subsection presents images of nutcrackers. There are several modern elephant nutcrackers - usually made out of wood and painted and decorated for the Christmas season. But there is at least one well-known, vintage iron nutcracker that is also shown.

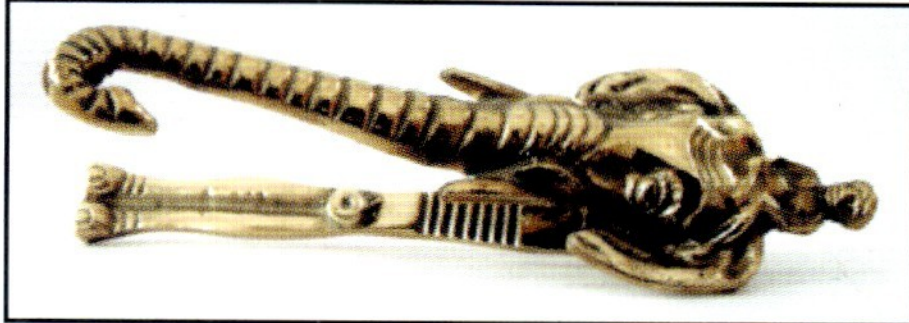

Nutcracker with "Item 3623" impressed on back. Made of cast iron with a gold-colored plate. A registration number of "712603" dates this to c. 1925. There are two cracking positions for large and smaller nuts. 7" H x 2" L. $25. *From the author's collection.*

Swank Nutcracker is a limited edition; number 3927 of 5000. 12" H. $25-35. *Courtesy of Frank Isbell.*

Cast iron art deco painted nutcracker. 5" H x 9.5" L. Weighs 4 pounds. 1950s. $75-125. *Courtesy of Don McKinney.*

Section 03.084 Nodders

This subsection presents images of elephants whose head is actually attached to the body with some mechanism like a spring, that enables the head to "nod" up and down or back and forth, hence the name "nodder." Another nodder is shown in the GOP/Political category. I know there are several older nodders in existence, mostly made from celluloid.

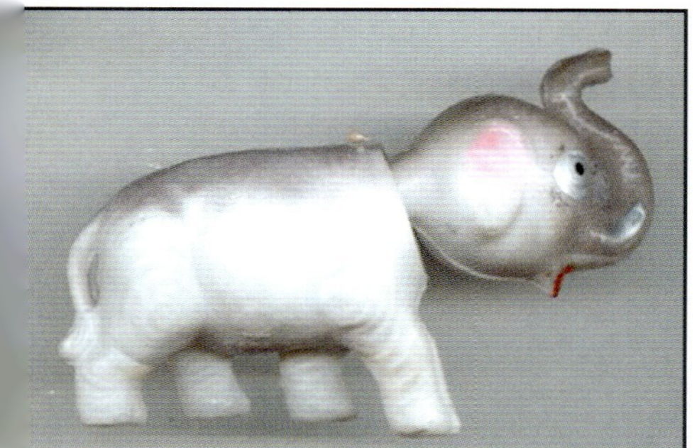

Thin celluloid hand-painted nodder. Stamped "Made in Japan" on his belly. 2" H x 4" L. 1940s-50s. $50. *Courtesy of Randy Haid.*

Composition nodder. 5.5" H. c. 1950. $30.

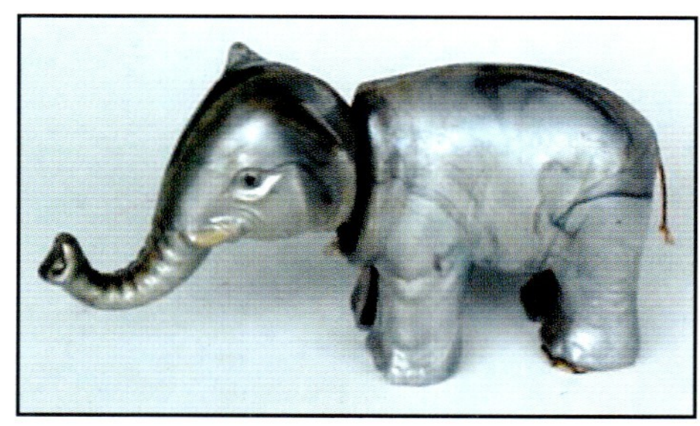

Gray hard plastic nodder with string tail. A sticker on the bottom of his foot is marked: "SA Reider and Co., NYC Copyright." Made in Breba, Germany. 3" L. c.1950. $40-50. *Courtesy of Jerry Lasseter.*

Section 03.086 Ornaments

This subsection presents images of ornaments. Some of the most recognized and sought-after of Christmas ornaments are those from Radko. Bugatti hood ornaments, both reproductions and the originals, are also popular, the originals commanding high prices.

Stuffed cloth Christmas ornament. 6" H x 3.75" L. c. 1960s. $5. *From the author's collection.*

Reproduction nickel-plated solid bronze 1931Type 41 Bugatti Royale hood ornament made by a Madrid, Spain jeweler. Only 6 Bugatti Royale automobiles were built and less than 10 original hood ornaments were ever created b the Valsuani foundry, the last of which sold at Christie's auction in 1999 for $35,000. This ornament was cast fror one of the 6 original Bugatti hood ornaments sculpted by Rembrandt Bugatti (brother of Ettori who built the Royale using the same lost wax method as the original. See *Car Collector, 1980*, for additional information. 8" H w/o stan c. 1970s. $550. A 6.5" H. silver-plated bronze, recently made radiator cap with an ornament of the same design valued at $200. Franklin Mint versions of this ornament a $150. *Courtesy of Walter Harris.*

Christopher Radko Christmas ornament made in Poland. 5" H. 1994. $40-60. *Courtesy of Martin DeFelice.*

Gilded metal ornament. 3.5" H x 5" L. 1990s. $7. *From the author's collection.*

Lladro porcelain Christmas ornament, model #6388 called "Circus Star." 3" H x 2.75" L. 1997; retired 1998. $50-100. *From the author's collection.*

Christopher Radko Christmas ornament. 5" H. 6" diameter. 1996. $40-60. *Courtesy of Martin DeFelice.*

Section 03.087 Paintings and Prints

This subsection presents images of paintings and prints. There are so many of these that it would take at least one separate volume just to present them all. Here I show several works that are representative of the art of capturing elephants on canvas or as limited edition prints. Some of the more famous artists in this category include Leroy Neiman, Gamini Ratnivira, Jiang, and Boulanger.

Graciela Rodo Bolanger print "Le Voyage Imaginaire" hand-signed and numbered limited edition of 750. 10" H x 9" L. 1993. $400. *From the author's collection.*

Print of elephant during the Prince of Wales's visit to Lahore India, from the *Illustrated London News.* 10" H x 16" L. 1906. $15. *From the author's collection.*

Print "Study of Elephant," black chalk, British Museum. Rembrandt, printed by F. Schmidt, Paris Printer. 1886. $40. *From the author's collection.*

Print by Angela Knipe, Wizard and Genius-Idealdecor AG, Printed in Switzerland. 7.5" H x 10" L. 1999. $25. *From the author's collection.*

An Indian zoomorphic print, a technique that makes one thing out of other things, in this case an elephant out of other animals. c. 1890. $850. *Courtesy of Elyahou and Helene Talasazan.*

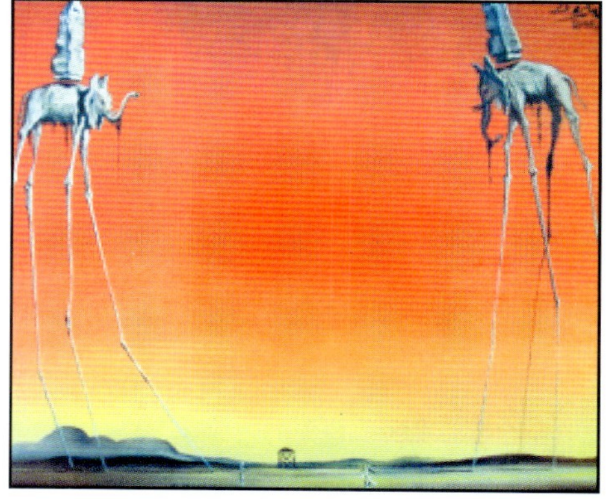

Dali Les Elephants Print, Descharnes, and Descharnes, Wizard and Genius-Ideal decor AG, Switzerland. Dali also created other works with elephants including: "Swans Reflecting Elephants" and "LaTentation." 24" H x 31.5" L. 1997. $25. *From the author's collection.*

Jiang Tie Feng's "Elephant Family" print. 40" H x 40" L. 1993. $6000. *Courtesy of Fingerhut Publishing.*

Miniature print done in conjunction with Leroy Neiman's classic serigraph, "Elephant Charge" by Knoedler Publishing. Neiman created several elephant paintings/prints include "Bull Elephant;" that serigraph is valued at $1400. Others include: "Circus," "Elephant Charge," "Elephant Family," "Elephant Nocturne," "Elephant Stampede," "Kilimanjaro Bulls," and "Shikar." You can view Neiman's elephant works at: www.leroyneiman.com. April 1999. $10-15. *Courtesy of Randy V. Jackson.* (E2D4).

"In Charge!" is a charging bull and lilac-breasted rollers oil on canvas by Gamini Ratnavira, a noted wildlife artist who has also designed elephant stamps for Sri Lanka and raised an elephant himself! This piece was donated to the Consulate of the Republic of Malawi, via Dr. James Clements for the international "Operation Malawi" fundraising event. Other works by Ratnavira can be seen at: www.gaminiratnavira.com. 20" H x 16" L. $5000. *Courtesy of Hidden Forest Art Gallery Fallbrook, California.* (E2D4).

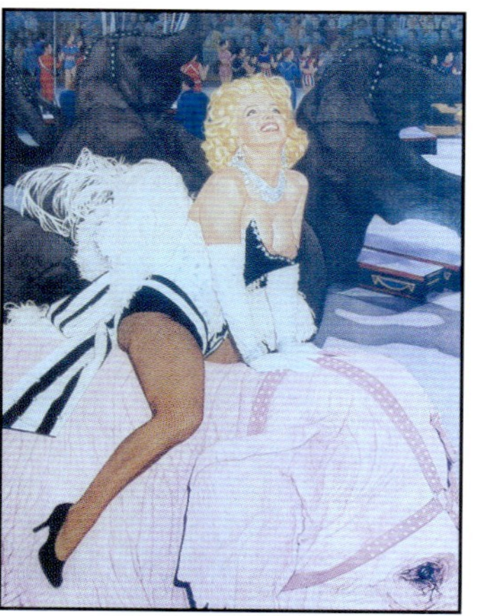

Print of African Elephant, by Henry J. Johnson in the book *Animal Kingdom*. 10" H x 6.5" L. 1880. $5. *From the author's collection.*

Classic woodcut engraving titled "Combat Between and Elephant and Rhinoceros" from a natural history volume by J.W. Buel. 1886. $10. *From the author's collection.*

Original oil-on-canvas painting titled: "There's No Business Like Show Business," by Cesear Vasallo. It is based on the black and white photo of Marilyn Monroe sitting on a pink elephant at a circus (see Photos section). 60" H x 45" L. 1998. *Courtesy of, without payment to, Cesear Vasallo.*

Section 03.088 Paperweights and Clips

This subsection presents images of paperweights and paper clips. Of course, many smaller, but relatively heavy elephant figurines could also be used as or called paperweights; but I included here only those things that I believe were intended to be paperweights by their maker.

Millefiori glass paperweight has small segments of differently shaped and colored glass rods pulled throughout a solid glass shape. In the "millefiori" technique, which means "a thousand flowers," rods are laid close together and then fused into tiny mosaics. The technique was first introduced by glassmakers from Murano, Italy. 2" H x 3.5" L. 1990s. $25. *Courtesy of Adele Verkamp*.

Rare paperweight from Cowan Pottery who manufactured from 1913-1931 in Rocky River, Ohio. This elephant has a gray-green glaze, is signed "COWAN" and has the embossed seal under the signature. 4.5" H x 3.5" L. 1920s. $225-300.

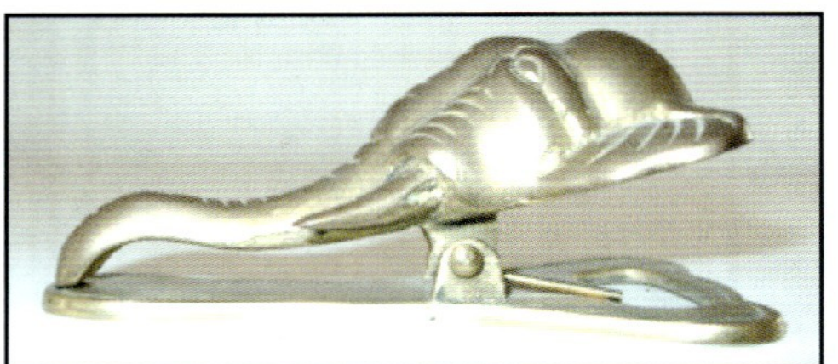

Large brass paper clip. 4.5" H x 2" L. 1980s. $15. *From the author's collection.*

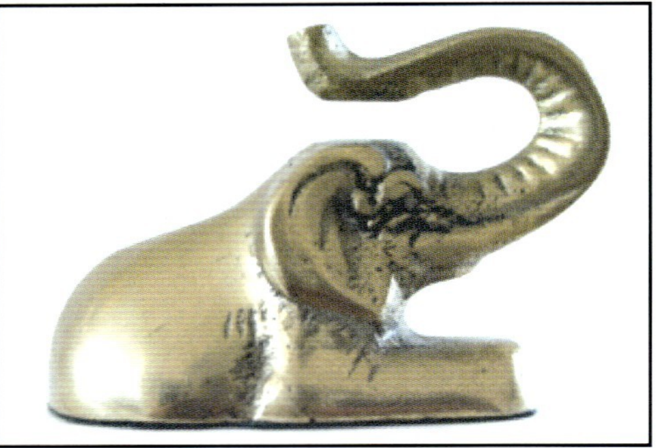

Brass paperweight. 1.75" H x 2.5" L. 1980s. $10. *From the author's collection.*

Rare Rookwood paperweight, model #6490 manufactured in Cincinnati, Ohio. The color is brown, with a hint of a green underglaze. On the underside is impressed the Rookwood trademark logo, Roman numerals XXXIV and 6490. In 1984, a limited edition 24Kt gold finish clone of this paperweight called Golden Elephant was produced after the original design by Kataro Shirayamadani. 3.75" H x 3.5" L. 1934. $225. *Courtesy of Richard Larose.*

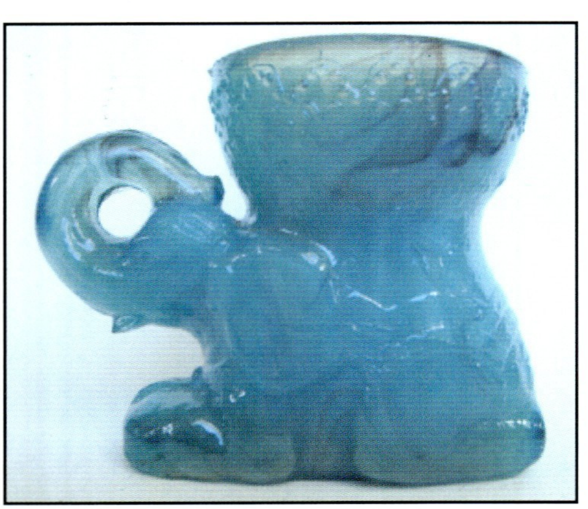

Confetti blue milk glass paperweight called Packy by Betty Craig of Craig's Glass Arts of Indiana. One of only 676 made. 2.5" H x 3.5" L. 1980. $15-20. *From the author's collection.*

Section 03.089 Pencils, Pens and Sharpeners

This subsection presents images of both writing implements and pencil sharpeners. One well-known pencil I own, but have not shown here, is a large-sized one with "The Elephant" and an elephant image on it.

Metal pencil sharpener made in Hong Kong. $10. *Courtesy of Harold Hubert.*

Ceramic pen with matching stand. There is a snake wrapped around a large pink stone on the pen. The pen is refillable with regular Papermate ink cartridges. 6" H x 3.125" diameter of the base. 1990s. $15.

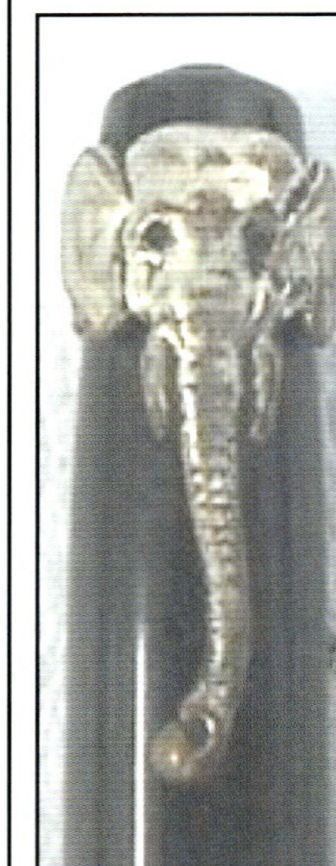

Waterman's fountain pen. On the side it reads: "Waterman's Reg. U.S. Pat. Off. Ideal Made in the United States of America," and on the nib: "Waterman's Ideal Account 14kt." It has an elephant head and trunk clip. 5" H. $25. *Courtesy of Merilyn.*

Wood pen/pencil/crayon holder with an assortment of elephant pens and pencils. 4.5" H x 5.5" L. Holder: $10, Pens/Pencils: $3-5 ea. *Courtesy of Adele Verkamp.*

Section 03.090 Perfume Bottles and Atomizers

This subsection presents images of perfume bottles. Some famous names associated with other elephant items, including Baccarat, Moser, Daum, Galle, and Lalique, also made perfume bottles, some of which are elephants. The most valuable one I know about ($3000) is a vaseline bottle with elephant finial from Czechoslovakia.

Czechoslovakian yellow glass perfume bottle and applicator signed "MADE IN CZECHOSLOVAKIA." The applicator is an elephant. 5.25" H. 1950s. $300. *Courtesy of Dominic Buchta.*

Glass perfume bottle with metal front and cap. 2" H x 2" L. 1998. $30. *Courtesy of Janet Wojciechowski.*

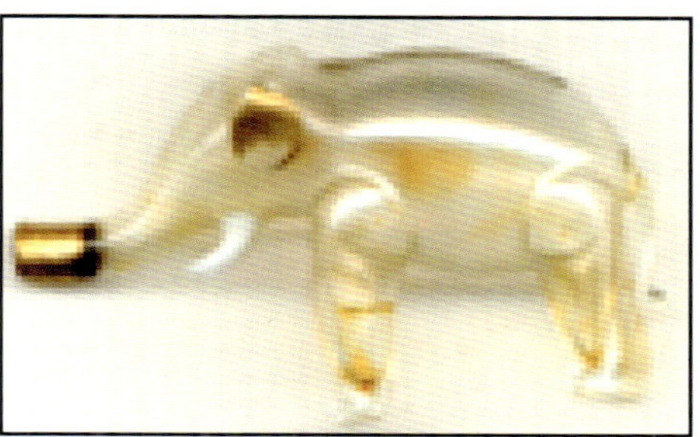

Small hand-blown glass perfume bottle similar to those made in Germany in the 1930s. Legs and tail usually filled with mercury for weight. $15-30.

Perfume bottle that, according to *Jones North, 1999*, had a paper tag reading "S. Kleinkramer Bergen op Zoom Holland." 1940s. $20. *Courtesy of Deborah K. Fisher.*

Section 03.091 Pez Dispensers

This subsection presents images of Pez dispensers and other Pez paraphernalia. Pez is a trade name for the dispensers and candy that came in them. Dispensers were made of plastic and were often crowned with a particular figure, including an elephant head.

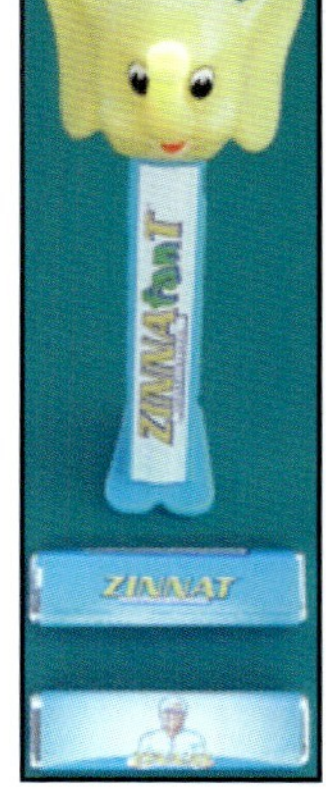

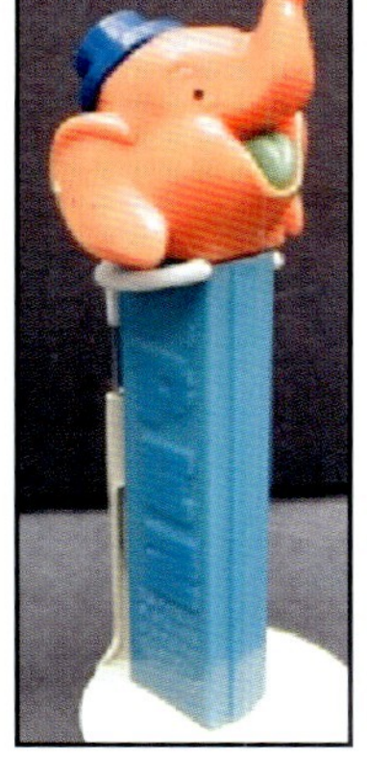

Blue with yellow head Pez. IMC 5 stem with sticker. Comes with one pack of PEZ candy. One side says ZINNNAT, the other side has a picture of a man holding the word zinnat. "Das Prktische Antibiotikum" is written on the side. Other side says GlaxoWellcome. 4" H. c. 1999. $10-15. *Courtesy of Tanya Wilson.*

Big Top Elephant "no feet" Pez marked "U.S. Patent 3,410,455 Made In Austria" on the side of the stem. The head pulls back and snaps back. His head is orange with a green tongue and dark blue hat. 4" H. 1970s. $70-85. *Courtesy of Andy Kurzadkowski.*

Section 03.092 Phone Cards

This subsection presents images of elephants that telecommunications companies placed onto phone cards.

Mita phone card. 2.25" H x 3.5" L. $10. *From the author's collection.*

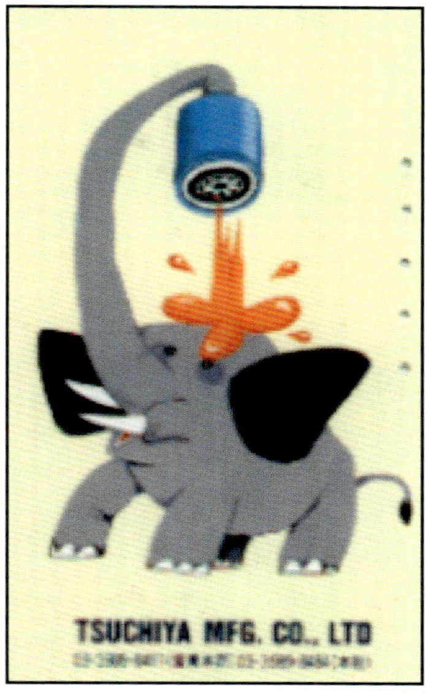

Tsuchiya Mfg. phone card. 2.25" H x 3.5" L. $10. *From the author's collection.*

Section 03.093 Photos

This subsection presents images of photos of actual live elephants.

Photo of Sinclair Inc. banner on an elephant. c. 1900. $20. *Courtesy of Scott Cross.*

Photo of Jumbo the elephant, who spent the first two decades of his life in captivity at England's London Zoo giving rides to thousands of children. He was sold to and toured with P. T. Barnum and Bailey Circus from 1882 to 1885 when he was killed by a train. The accident is captured in a die-cut, the image of which is shown the Die-cut category. 11" H x 14" L. 1880s. $20. *From the author's collection.*

Photo of the state elephant of the Gaikwar of Baroda. This colored photo depicts the Gaikwar of Baroda, one of the few independent native princes of India. He ruled over a district of nearly 5000 square miles in the Bombay Presidency and traced the origin of his power to the early part of the eighteenth century. Like all Hindu rulers and high officials in India, this potentate had a special state elephant, which was adorned with magnificent trappings and surmounted by a howdah, in which rode the prince and those whom he chiefly honored. The plate is cut from the *History of India*, edited by Abraham Valentine Williams (1862-1937), a limited edition published around 1906-07 by the Grolier Society. Plate: 6.75" x 9.75". Image: 4.5" x 6.25". c.1906. $15.

Black and white photo of Marilyn Monroe sitting on a pink elephant. It is from an article about pink elephants and purple hair in *Playboy*, September 1955, Volume 2 Number 9. The photo was taken on opening night of the Ringling Brothers circus at Madison Square Garden for benefit of the Arthritis and Rheumatism Foundation, on March 31st, 1955. $50. *Courtesy of, without payment to, Ceasar Vasallo.*

Section 03.094 Picture Frames

This subsection presents images of picture frames that have elephants on them.

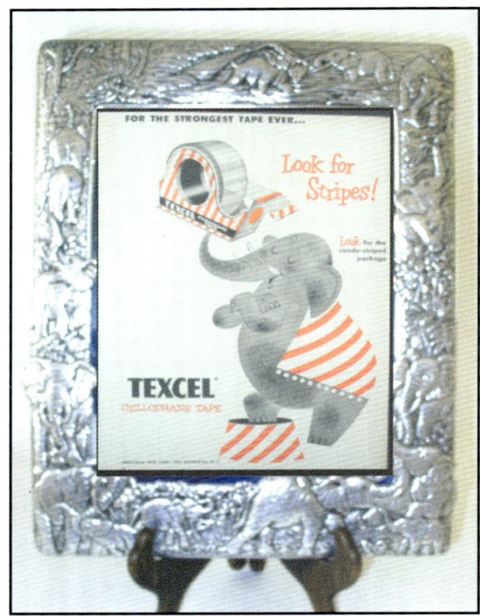

Aluminum picture frame with elephants all around the edge. 14" H x 11" L. 1980s. $25. The image inside the frame is a 1953 advertisement for Texcel Tape, Industrial Tape Co. *From the author's collection.*

Enesco brass picture frame, made in Taiwan. 3.25" H x 3.5" L. 1983. $10. *From the author's collection.*

Silver-plated picture frame. 2.5" H x 3" L. 1980s. $30. *Courtesy of Adele Verkamp.*

Section 03.095 Pie birds and Pie Vents

This subsection presents images of pie-vents, made mostly of heat-resistant materials (e.g., ceramics), some of which were made as elephant figures. You put them in the middle of a pie to vent steam, keep the pie from boiling over and to hold the piecrust out of the pie juice so it won't get soggy. They were invented in England in the early 1800s. The Nutbrown Pottery Company of England is a popular one to collect, as are ones from Cardinal China Company, which can range from $200-300 (for the swirl-base variety).

These elephants are pie birds and made by the Nutbrown Pottery Co. in England, in business from 1927 to 1988. The elephants are marked on the inside and have registration numbers. The white Nutbrown elephant is pretty common, but the gray one is very rare. 3.5" H. $70 and $150. *Courtesy of Lorraine Miller.*

Gray pie bird by Nutbrown of England. See *Collectibles for the Kitchen, Bath and Beyond, p. 37.* These Nutbrown pie bird values vary by date made, which can be determined by the patent number found on the inside. This one is "Patent 800828" and you can get the date from the British Patent office. 3.5" H. c. 1920s-1940s. $150-250. *Courtesy of Ellen and Marty Bercovici*

Pie bird made by "The Heart Of Evelyn," a potter in Paris Landing, Tennessee. Vented through the back of his head and notched on the bottom. 3.5" H. 1998. $20. *Courtesy of Lorraine Miller.*

Section 03.096 Pillow Covers

This subsection presents images of pillow covers that have an elephant theme.

Elephant pillow pattern. 22" H x 16" L. 1960s. $10. *From the author's collection.*

Pillow by Waverly, The Luxury Colle Pillow, polyester, needlepoint on chamois. 17" H x 17" L. 2000. $50. *Courtesy of the Great Indoors.*

Indian pillow cover. 15" H x 15" L. 1960s. $10. *From the author's collection.*

"Oricing" pillow by Judy Davis, Olallawa Pillow. 11" H x 14" L. $15. *Courtesy of Adele Verkamp.*

Section 03.097 Pink Elephants

This subsection presents images of a 1950s phenomena called Pink Elephants. Pink elephants were used to decorate several kitchen and drinking-related items, and I tried to show several things currently sought after. Manufacturers of these items include Cory and Hazel Atlas.

Pink elephant cocktail tip tray. 4.75" H x 6.75" L. 1950s. $7. *From the author's collection.*

A Guide to Pink Elephants bar mix guide, Richard Rosen Assoc. 4" H x 4.5" L. 1952. $12. *From the author's collection.*

Pink elephant glass ice bucket. 4.5" H x 5.5" L. 1950s. $40-50. *Courtesy of Adele Verkamp.*

Pink elephant cocktail recipe tin holder. It has 2 elephants on lid, 3 trunk-to-tail elephants on front, and 2 on each side. Bottom reads "Mayfair Co. A Mayfair Creation Chicago 6, Ill." 3.5" H x 5.25" L. c.1950s. $35. *Courtesy of Mike.*

Pink elephant glass carafe, probably by Cory. 5" H x 4" L. 1950s. $25. *Courtesy of Adele Verkamp.*

Pink elephant juice glass. 3.25" H x 2.5" D. 1950s. $6. *From the author's collection.*

Pink elephant "Kool Kubes," reusable Ice Cubes. 1950s. $10. *From the author's collection.*

Pink elephant cocktail shaker with aluminum cap. 1950s. $40-50. *Courtesy of Bernard Poda.*

Hazel Atlas pink elephant glass tumblers from their glass plant in Clarksburg, West Virginia. Unsigned Zombie tumbler: 6" H.; signed Juice tumbler: 4.875" H. c.1950s. $15 and $10. *Courtesy of Briar Patch Antiques, Clarksburg, West Virginia.*

Section 03.098 Pins

This subsection presents images of pins or slogan-based buttons (not the kind that keeps clothes together, but the kind that you wear on clothes to express a sentiment). There are perhaps thousands of GOP-Political pins with an elephant on them - some of which are in that category.

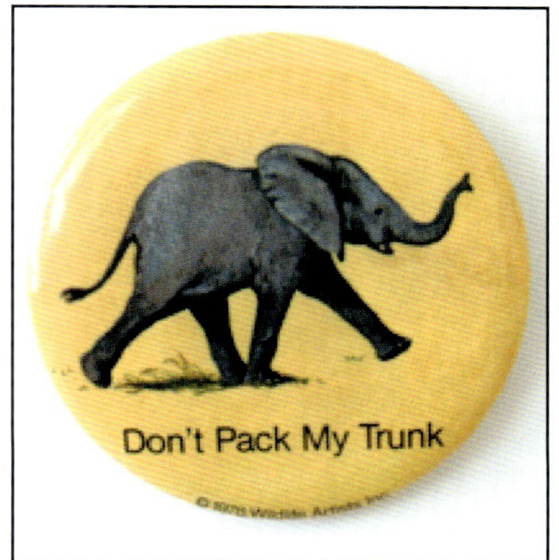

Wildlife Artist Inc. pin that says: "Don't Pack My Trunk." 2.375" D. 1978. $5. *From the author's collection.*

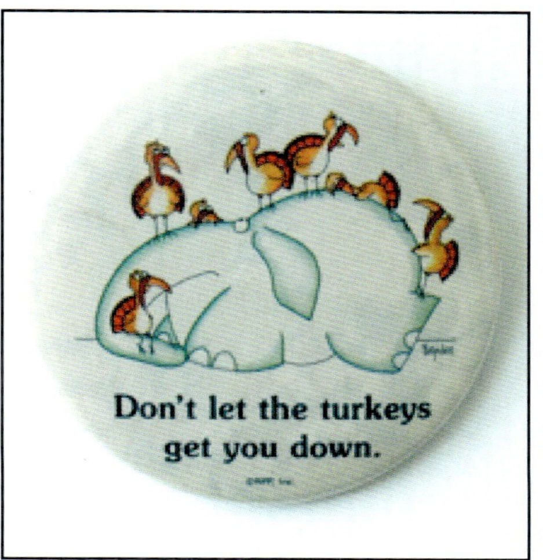

RPP Inc. pin that says: "Don't let the turkeys get you down." 2.25" D. $3. *From the author's collection.*

Elephant ashtray and pipe stand with integral match holder. The ashtray insert is glass and the stand has a marble base. Bottom is inscribed: "555." 28" H. c.1930s-40s. $200. *Courtesy of John.*

Section 03.099 Pipe Holders

This subsection presents images of pipe holders that have an accompanying elephant figure holding up the holder or decorating it in a "side-car"-like arrangement.

Victorian brass pipe holder. 3.25" H x 5.5" L. c. 1890-1910. $40. *From the author's collection.*

Pewter pipe holder, made in Japan. 2.5" H x 4" L. c. 1990. $20. *From the author's collection.*

Pewter pipe tamper; #22 in a series by Catnip Hill. 2.25" H x 1.25" L. c. 1990s. $20. *From the author's collection.*

Oak dual pipe holder. 4.5" H x 9" L. c. 1980. $15. *From the author's collection.*

Section 03.100 Pipes

This subsection presents images of pipes that have been made as an elephant head or have an elephant image carved into them. Meerschaum (a clay found in Turkey) is a favorite pipe material, and I have seen several elephant-head meerschaum pipes. Of course, metal and ceramic pipes have been made with some sort of elephant theme as well, if only to decorate the stem.

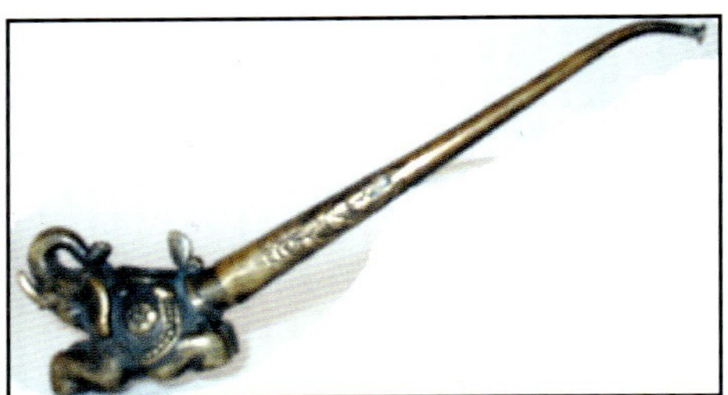

Chinese brass pipe. 9.75" H x 1.5" L. $25. *From the author's collection.*

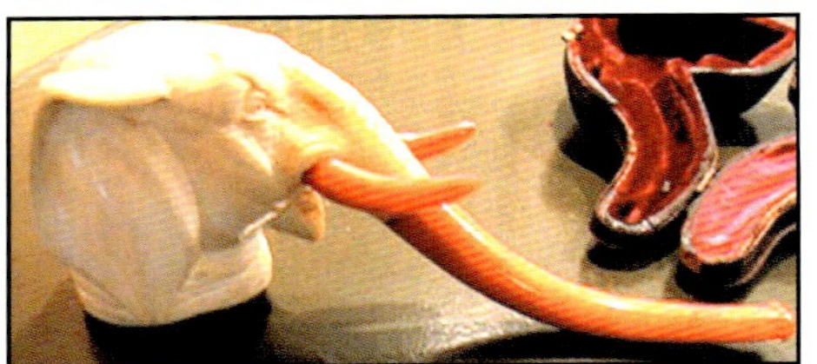

Meerschaum pipe with amber tusks and the trunk, which is the stem, is amber as well. 6.5" L; Bowl is 2.75" H. 1800s. $1000. *Courtesy of Cory Margolis.* (E2D4).

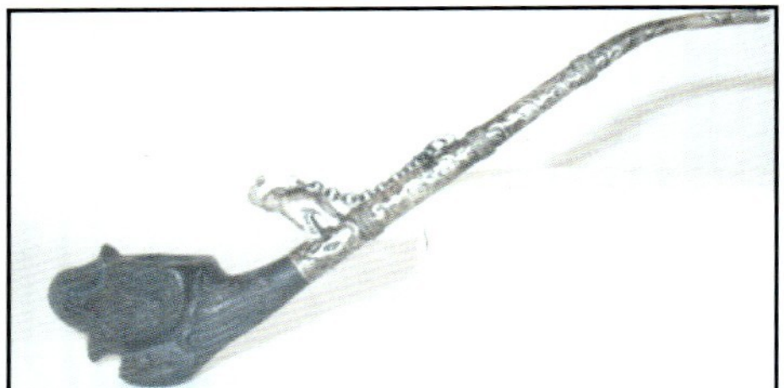

Chinese pipe made of fired clay bowl and silver-plated metal stem with small elephant attached and chained to stem. 10.5" H x 1.25" L. $20. *From the author's collection.*

Meerschaum pipe with elephant relief carved into bowl. Stem is made of Lucite. The pipe is listed in the CeSem catalog as Collector Series CS-060 (Elephant). 6.25" L; 2.125" H. $75-100. *Courtesy of Tom of CeSem-america.com.*

Section 03.101 Pitchers, Decanters and Creamers

This subsection presents images of containers for storing and pouring liquids; teapots fit this definition but are in their own category.

Fitz and Floyd ceramic pitcher. 9" H x 9" L. 1985. $40-50. *Courtesy of Adele Verkamp.*

Sarreguemines ironstone pitcher ornamented with a gray cama'eu medallion with elephants in an Indian landscape. Frame and borders are decorated with lustre transfers. Signed on bottom. 10" H. c.1875. $100-150. *Courtesy of eclectic-antiques.com.*

Staffordshire pitcher with turquoise glaze at top shading to brown at the bottom. It is stamped on the bottom with a phoenix bird rising out of flames and the letters "T.F.andS.L. England." According to *Kovel's New Dictionary of Marks* it was made by the Thomas Forrester and Sons Ltd. Company in Longton Staffordshire, England. The company was in business from 1883 to 1959. 11" H. $375. *Courtesy of Kathy Heinowski.*

Ceramic creamer made by Shawnee Pottery was designed for a child's feeding set by Rolf J. Falk, Zanesville, Ohio in 1946. He is a creamy white with burgundy color in the tip of his ears and mouth. His eyes and toenails are brown, and his tusks are gray. It is marked "Patented U.S.A." on the bottom and has a red Shawnee sticker on the side; several models of this shape were made and the value depends on the style and color of painted details. 4.25" H x 5" L. 1946. $50. *From the author's collection.*

Majolica jug made in France by Sarreguemines. Signed on the bottom: "SARREGUEMINES DIGOIN 4470." 9.5" H. c. 1890s. $400-500. *Courtesy of Mr. Farnault.*

Goebel porcelain creamer with yellow body, white and black eyes and handle, and black feet. Marks on the bottom are as follows: incised crown and "TMK-1," number "187" the usual "O 1/2" incised size code, letter "S," letters "DEP" incised below the crown, "GERMANY" stamped in black, and the lett "M," which is the artist's mark. There is a faint stamped mark to the left of "DEP" that looks like a "W" along with somethin else. 5" H. c.1920s. $50. *Courtesy of Sharon Murray.*

Section 03.102 Planters

This subsection presents images of the many elephant planters out there. The most sought-after names include: Shawnee, Royal Copley, Red Wing, Royal Haeger, and Rosemeade.

Mann ceramic planter, made in Japan. 7" H x 10" L. $25. *From the author's collection.*

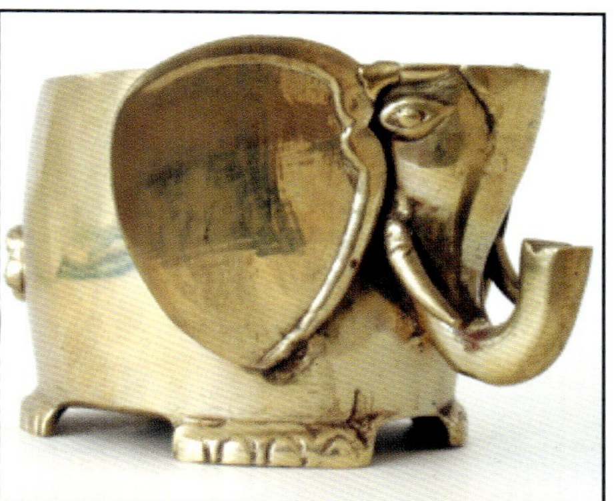

Brass planter, made in India. 3" H x 6" L. $25. *From the author's collection.*

Unmarked Shawnee planter. 5.5" H x 7.5" L. 1940s. $15-35. *From the author's collection.*

Ceramic planter with original paint by Hull of Crooksville, Ohio from 1905-1986. Other colors included green, yellow, off-white/ cream, and blue. 5.5" H x 5.25" L. 1950s. $30-75. *From the author's collection.*

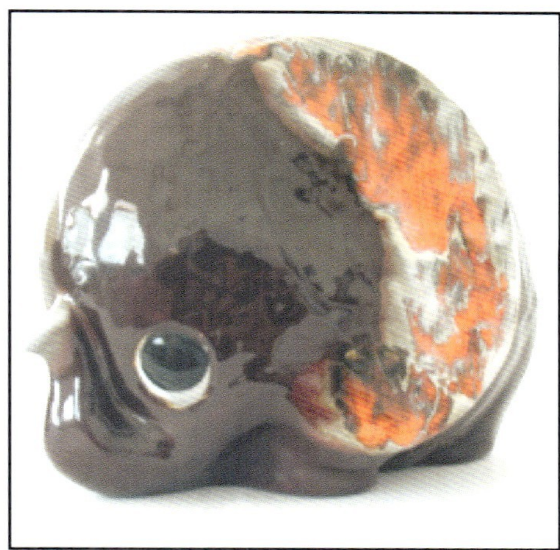

Claylick ceramic planter. 6" H x 8" L. $15. *From the author's collection.*

Brad Keeler Pottery planter. In 1946, after a fire completely destroyed the American Pottery Plant, Keeler moved his studio and began producing planters and figures of children, animals, and baby-related articles known as, "PRYDE and JOY." This hand stitched-looking elephant has a paper label on the belly reads, "HAND DECORATED PRYDE and JOY LOS ANGELES, CALIF." 5.5" H x 6" L. 1946. $35-50.

Green ceramic lamp base planter with rider by Royal Haeger, model number R-563. 11" H x 10.5" L. c. 1950s. $75-125. *From the author's collection.*

Shawnee Planter, marked USA. 4.5" H x 3.75" L. 1940s. $25. There is a 3" H model too. *From the author's collection.*

Haeger elephant pottery planter. Marked on bottom "Haeger, USA." 5" H x 12" L. 1943. $50. *Courtesy of Theresa Saunders.*

Royal Copley elephant on ball planter by Spaulding China Company of Sebring, Ohio, which made pottery from 1939-1960. 7.75" H. 1950s. $40. *Courtesy of Lana G. Connolly.*

Shawnee planter from Zanesville, Ohio from 1937-61. 6" H x 6" L. 1950s. $50-75.

Red Mill elephant planter #754 made of crushed pecan shells. 10" H x 10" L. c. 1990s. $50. *Courtesy of Deb.*

Section 03.103 Plates and Trays

This subsection presents images of plates, dishes and trays. Many of the major dinnerware manufacturers have made patterns that include an elephant, Dedham being a standout. Some of the most sought after pieces are special commemorative plates. There are additional plates in the Political/GOP section.

Brass with silver wash and copper inlaid plate; has central elephant and was made in India. 18" D. c. 1800s. $150-200. *From the author's collection.*

Japanese porcelain and brass plate hand-decorated in Hong Kong for Trifles. 8" D. $50. *From the author's collection.*

Bradex plate "Curiosity Asian Elephant" from W. S. George Fine China. 1992. $30. *Courtesy of Adele Verkamp.*

Emal de Limoges 1. Godinger plate. 10" D. 1855. $75. *Courtesy of Adele Verkamp.*

Carlsberg Malt Liquor plastic tray. 11" H x 15" L. $15. *From the author's collection.*

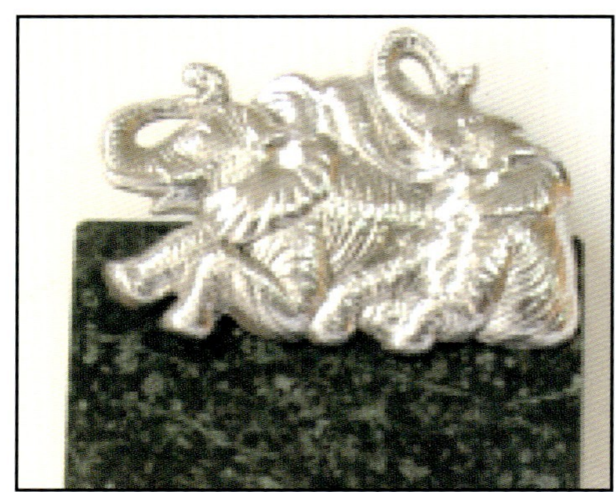

Cutting board from Arthur Court made of granite with aluminum elephant motif at one end. 20" H x 6" L. 1987. $75. *Courtesy of Adele Verkamp.*

China Elpco plate, USA. 6.25" D. $50. *Courtesy of Adele Verkamp.*

Dedham Pottery elephant plate in dark blue with finely spaced uniform crackle. This example is unusual without the baby elephant that is seen on Dedham plates. It carries the registered stamp mark and two impressed rabbit marks. Dedham Pottery was produced by the Robertson family in Dedham, Massachusetts between 1896 and 1943. Reproductions of various works are produced by: Dedham Historical Society, the Starr family in West Concord, Massachusetts, and the Nashawaty family in Walpole, Massachusetts. Note that a Centennial 1997 plate was made much like this. 6.375" diameter. 1929-1943. $1000. *Courtesy of Jim Kaufman* (E2D4).

Ceramic shot glass tray with 4 shot cups made in Japan. The side says "Whiskey." 6" H x 7" L. pre-1938. $75. *Courtesy of Janet Wojciechowski.*

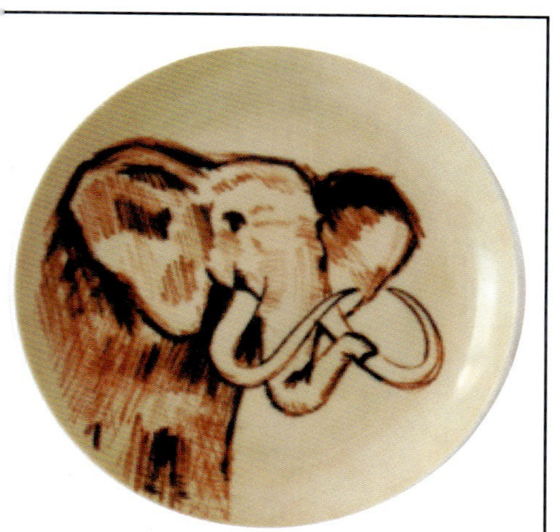

Fitz and Floyd plate with "FF Japan" Sticker. 8" D. 1976. $25. *From the author's collection.*

Fremlins Beer advertising tray. The tray is made of tin and advertises Fremlins Brewery, which brewed beer from 1928 until it was taken over by the Whitbread company in 1967. 13.5" H x 13.5" L. 1960s. $15-30. *Courtesy of Mary Marwick.*

Cobalt glass cup plate with scalloped edge by Pairpoint Crystal Company. 3.25" D. 1992. $25. *From the author's collection.*

Porcelain Bing and Grondahl Mother's Day plate made in Copenhagen, Denmark. Marked " #9386." 5.875" D. 1986. $35. *From the author's collection.*

Ash plate. 7" H x 6" L. 1995. $25. *Courtesy of Janet Wojciechowski*

Section 03.104
Playing Cards and Accessories

This subsection presents images of elephant playing cards and card-playing accessories. Another deck I know of is Fifth Avenue Cards.

"Olifant Speel Kaarten" playing cards made by Carta Mundi in Belgium. 3.5" H x 2.25" L. 1970s. $20. *From the author's collection.*

Four painted and numbered ("1" to "4") iron elephants connected by a tasseled string (running through the elephants' noses) to silver pencils. These are used as bridge accessories. They are not marked. Elephants: 2" H. pencils: 2.5" L. Late 1930s or early 40s. $50-75.

Tier-Quartett playing cards. Quartett is a German game, similar to "Go Fish." The idea is to collect a whole "suit" of animals of a particular type, of which there are 9, with four cards in each. The cards are made by Berliner Spielkarten of Germany, a well-known company still in existence. The cards are stored in a clear hard plastic box, which has the elephant "title" card (the image on the right) placed on top of the deck to identify what's contained. The company had a contract with Steiff to make this deck, as the elephant shown is a Steiff elephant; there are elephant cards in the deck as well. 4" H x 2.625" L. c.1950s-60s. $25. *Courtesy of Rosalie Isaacs.*

Section 03.105 Poker Chips

This subsection presents images of poker chips with elephant images. I could have put this in the Playing Cards and Accessories category, but there are several of these and not all card games use chips. The earliest chips I found date to the late 1800s. Elephant images are usually engraved or inked onto an ivory, clay, wood, or plastic substrate. Composition chips were developed in about 1880 to replace the ivory chips of the day. Ivory chips are probably anathema to elephant collectors, but most American ivory chips are from whale (which may still not be a comfort). Generally, certain colors are more valuable than others, white being the most common, red and blue come next, then yellow or any other color.

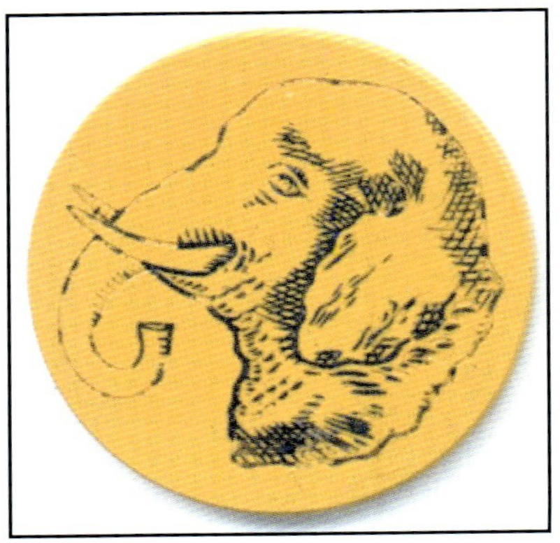

Clay composition poker chip, white and quite rare. 1.5" D. $50-80. *Courtesy of Michael Par.*

Clay composition poker chip; known as Lucky Elephant. It is Seymour Code PA-GE. 1.5" D. 1920s-30s. $1-3. *Courtesy of Michael Par.*

Clay composition poker chip, white, referred to as "Tied Elephant." 1.5" D. c. 1900-1905. $15-30. *Courtesy of Michael Par.*

A crude clay composition poker chip. 1.5" D. c. 1880s. $20-40. *Courtesy of Michael Par.*

Clay composition poker chip; one of the earliest clay chips. 1.5" D. c. 1880. $4-8. *Courtesy of Michael Par.*

A finely engraved chip, this one was never inked. 1.5" D. c. 1886. $20-30. *Courtesy of Michael Par.*

Section 03.106 Political and GOP

This subsection presents images of different types of elephant items that relate to the Grand Old Party (GOP); Thomas Nast created its mascot, the elephant. Over the years since Nast first "assigned" the elephant to the GOP (and the ass to the Democrats - ha ha), there have been millions of items made that would fit this category. National, state, and local elections, every election year, produced buttons, badges, hats, banners, cups, plates, ties, etc., to commemorate candidates or issues.

Republican National Committee Nixon bronze token in hard plastic case with information card, made by the Franklin Mint. 1.5" D. 1972. $10. *From the author's collection.*

Republican campaign decanter produced by Wheaton Glass (New Jersey). It is an amber color; amethyst also exists. Embossed image of Nixon on one side, Agnew on the other with words: "For President" and "For Vice President" below. "REPUBLICAN" embossed on base of Nixon side, "CAMPAIGN" embossed on base of Agnew side. 7.25" H x 6" L. 1972. $30. *From the author's collection.*

Plaster "Nixon For President" nodder, made in Japan. 7" H x 4" L. c. 1972. $25. *From the author's collection.*

Jim Beam Republican flask with elephant holding up world a la "Superman." 11.5" H. 1980. $60-85. *Courtesy of Huck Leonard.*

Porcelain elephant with "GOP" on back blanket. 3.75" H x 4.5" L. 1984. $15. *From the author's collection.*

(Alf) Landon - (Frank) Knox celluloid pin. .875" diameter. 1936. $7-10. *Courtesy of Brian Sample.*

Cigarette tin with "GOP" and "1948" painted on. $10.

Cast Iron elephant with gold-painted raised letters on the painted blanket area read: "Land on Roosevelt 1938." 3" H x 5" L. 1938. $360. *Courtesy of Rebecca.*

Cobalt pottery elephant with raised trunk and letters on one side reading: "AIKEN" and on the other side "GOP." It was given out in Vermont when Aiken ran for the senate; he served from 1940 through 1962. The elephant is made by Morton Pottery from their model #606 elephant figurine. This caught on with other politicians including Richard M. Nixon, and the elephants made great campaign giveaways from the 1940s through the 1950s. 2.5" H. c.1940-1960. $30. *From the author's collection.*

Frankoma GOP mug, US Pat D215-868. 4" H x 5" L. 1976. $30-40. Many of these mugs were made: an elephant every year since 1968, and, since 1975, an elephant and donkey (the Democratic mascot). The rare 1974 Nixon/Ford mug in coffee glaze is worth $400-600, as only 400-500 were produced. The 1996 white glaze elephant mug is worth between $65 and $80. *From the author's collection.*

Pewter figurine on plastic base campaigner, with "Balfour Jewelry's Finest Craftsmen" sticker. 3" H x 4" L. 1972. $25. *From the author's collection.*

Two "G.O.P." Republican Party pins from the turn of the 20th century. The one to the left has a buttonhole stud rather than a pin. The pin on the right with the blue elephant and "Grand Old Party" on top has its original back paper intact from the General Mfg. Co. of 162 Fulton St. in New York City with patent dates listed of: July 17, 1894 and April 14, 1896. It is from the 1896 presidential election because the Republican theme for that election was the gold standard and sound money, and many of the campaign items were gold colored to fit that theme. Has: "GOOD AS GOLD" at the bottom of the pin. 1919 and c.1896. $15 and $20. *Courtesy of Bill Darcy.*

Celluloid Young Republican pinback button. 1.5" diameter. c.1930-1940s. $10.

Celluloid button for Parker and Davis; multi-slogan, football-related pinback has White Elephant GOP Recapitulation back paper. 1.5" diameter. 1904. $400-500. *Courtesy of Brian Sample.*

GOP Everett McKinley Dirksen ceramic ashtray. This is an Illinois campaign memento by Morton Pottery. The following information is according to Doris and Burdell Hall: "Illinois Senator Everett McKinley Dirksen was one of the company's best customers. His campaign headquarters around the state handed out hundreds of the little elephants. When Dirksen campaigned for the House of Representatives in 1944 the pottery designed a special ashtray to be handed out to delegates at the Republican convention in Chicago. Octagonal in shape, and glazed in Delft blue, the ashtray has this inscription across the top rim: 'Everett McKinley Dirksen.' Around the bottom rim is: 'From Lincoln's Illinois' In the bottom of the ashtray is an elephant with "GOP" above its back. Twenty-five hundred ashtrays were distributed at that convention by Mortonites who were friends and political hackneys of Dirksen." 3.875" L. 1944. $20. *From the author's collection.*

Hand-held fan with the Republican elephant sitting on a donkey. The eyes on the elephant are applied and when you move the fan, it winks at you. It is a Barry Goldwater Fan Club item and was made by Artcraft. 10.5" H x 6.25" L. 1960s. $20. *Courtesy of Barbara.*

GOPeanut Crunch tin manufactured by the Earle S. Bowers Co., Division of Chas. J. Webb Sons Co. Inc., Philadelphia, Pennsylvania. GOPeanut Crunch was sold to raise money for the Republican Party. Banner reads "Invest in the Future-Vote Republican." 15 oz. can. 5.5" H. 1963. $25. *Courtesy of Janet.*

Forty-watt light bulb with a GOP/elephant filament. c. 1940s. $30. *Courtesy of Bill Miller.*

"Win with Wilkie" hat from Chicago Cook County, Illinois. Republican Wendell Wilkie failed to unseat Roosevelt in the 1940 Presidential election. 1.5" H. c. 1940. $40. *From the author's collection.*

Section 03.107 Postcards

This subsection presents images of postcards. Most zoos or other locations that have elephants (like museums or the Elephant Hotel shown in one of the images), or an elephant theme, usually have elephant postcards. So there are many elephant postcards around dating to the late 1800s.

Jumbo Fruit postcard showing image from fruit label. 3.5" H x 5" L. 1980s. $5. *From the author's collection.*

Leather postcard with words: "I can get my meals by putting my trunk up;" an Arlington, South Dakota souvenir. 1905. $15. *From the author's collection.*

Elephant Hotel in Atlantic City, New Jersey linen postcard, by E. C. Krupt Co. 1955. $7. *From the author's collection.*

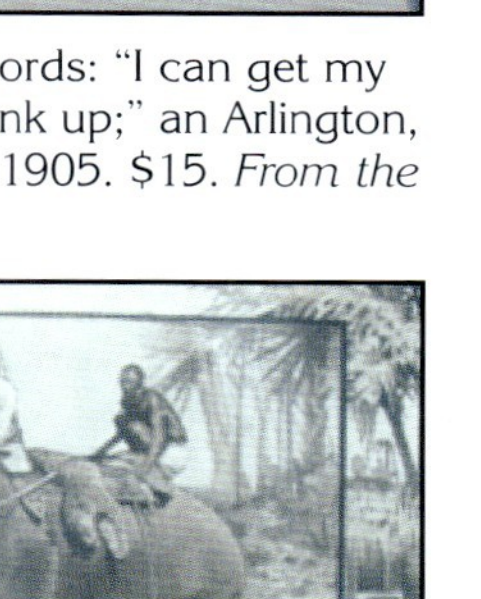

Postcard made by M.J. Mintz shows an African elephant helping with work. 1909. $5. *From the author's collection.*

Smithsonian postcard by Primere Litho showing largest elephant on record. 1955. $7. *From the author's collection.*

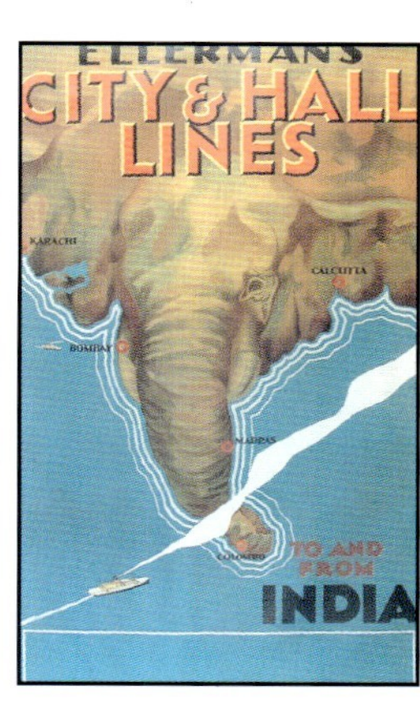

Postcard advertising Ellerman's Shipping from India designed by Mayfair Postcards of London. 6" H x 4.25" L. $5. *From the author's collection.*

Section 03.108 Posters

This subsection presents images of posters. Most elephant posters were used as advertising for either circuses or products. Many older posters run into the thousands of dollars for originals, hence many collectors satisfy their quest by obtaining moderately priced reproductions.

"Jumbo The Children's Giant Pet" Barnum Bailey and Hutchinson Challenge poster advertising a challenge of $100,000 that Jumbo is the largest elephant in the world. 16" H x 10" L. $15. *From the author's collection.*

Poster reproduction advertising tires: "Pneu Imperforable MENJOU," by Contact Habillages. 28" H x 20" L. 1998. $25. *From the author's collection.*

Poster reproduction advertising "Banania Exquis Dejeuner Sucre," a French cereal by Contact Habillages. 28" H x 20" L. 1990s of original art in 1910. $20. *From the author's collection.*

Poster reproduction advertising The Siam Cement Co. by Bangkok Poster. Note that there is an elephant image on the barrel top the big elephant is holding. 27.5" H x 19" L. 1990s reproduction of original artwork by R. Wening created in the early 1900s. $40. *From the author's collection.*

IMAX Corporation movie poster advertising "Africa's Elephant Kingdom," a documentary by Discovery Channel Pictures. 36" H x 24" L. 1990s. $25. *From the author's collection.*

Original poster advertising the movie "Lord of the Jungle." 1955. $100. *Courtesy of John Diele.*

Original TWA travel poster advocating consumers to fly to India by David Klein. 40" H x 25" L. 1950s. $200. *Courtesy of Karen.* (E2D4).

A famous French cigarette poster reproduction: "Je Ne Fume Que Le Nil" by Cappiello, this image can be found on a mini-poster, small tin sign, and the cigarette papers' wrapping as well. 47" H x 63" L. 1999. Original artwork produced c. 1906. $50. An original c. 1925 poster (same size) is valued at $750-1000. *From the author's collection.* (E2D4)

Section 03.109 Printing Blocks and Rubber Stamps

This subsection presents images of printing equipment that transfers an elephant image onto a substrate like paper or wax, or that uses an elephant as the design motif for a stamping device. I have seen several nice Chinese bronze seals with an elephant as a handle.

Arizona Stamps Too inkstamp model #G15. 4" H x 5" L. 1990s. $15. *From the author's collection.*

Three inkstamps: GOP Centennial on left and 2 others are all metal-on-wood construction. 2" H x 2" L. 1" H x 1" L. 1.75" H x 1.75" L. $5, $3, and $10. *From the author's collection.*

Chinese bronze stamp. 1800s. $50-75. *Courtesy of Ron Lukas.*

British Empire iron die. It has the lettering "Dumbarton Reserve" around the edge. There is an image of an elephant carrying a howdah, all under a crowned crest. These historic pieces were imported over a quarter of a century ago from the foundry in England where badges, medals, decorations, and insignias were created from these dies. 1.75" H x 1.625" L x .25" D; weighs 3.8 oz. 19th and early 20th century. $30-60.

Copper-faced print block used for printing labels by the Oriental Powder Company of South Windham. During the U.S. Civil War the company was a major supplier of gunpowder to the U.S. Government. The elephant symbolizes India and the surrounding areas where the company imported their raw materials (nitrates) for gunpowder. The print block reads: "ORIENTAL POWDER COMPANY MANUFACTURES of GUNPOWDER, for sale at 160 State Street, Boston, Mass. 25 Exchange Street, Portland, Maine." 2.875" H x 4.375." L x .75" D; weighs : 3.8 oz. 1850-1900. $50-85. *Courtesy of Francis Raynes.*

Section 03.110 Purses

This subsection presents images of elephant purses; Judith Leiber produced the most sought after elephant purses, while Faberge made at least one silver purse with an inscribed elephant.

Black vinyl purse with gold colored metal elephant decoration, made by NIMA, China. 8" H x 6" L. 1990s. $40. *Courtesy of Adele Verkamp.*

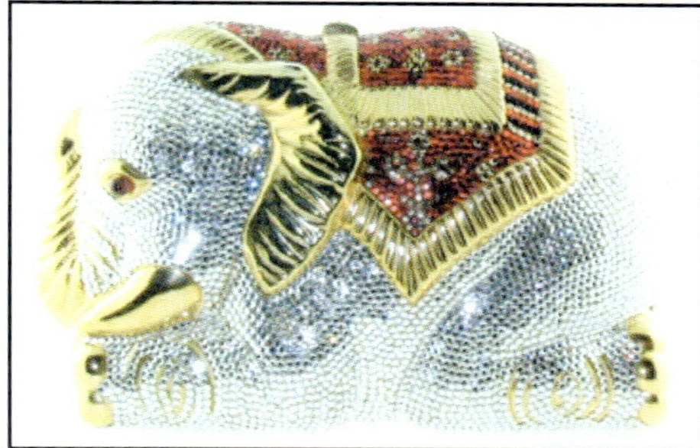

Judith Leiber evening bag, also called a minaudiere in French, is completely encrusted with Austrian crystal rhinestones. Sculptured trunk and feet, "elbows," and ears. The saddle area is especially notable for its mosaic of design in crystal. In 1963, after working 13 years in the American handbag industry, Judith Leiber launched her own firm creating the first minaudiere in 1967 and they are still sold today. 4" H x 5.5" L. 1997. $2500 half-beaded, $3800 full bead. *Courtesy of Marvin Eisen of Estates On Line.*

Celluloid purse frame with carved elephant designs and elephant clasp. 1920. $75.

Section 03.111 Puzzles

This subsection presents images of elephant puzzles.

Wood puzzle signed "D.D." 9" H x 8.5" L. 1985. $25. *Courtesy of Adele Verkamp.*

Wood puzzle made in China, #WB307 with 9 pieces. 6" H x 7.5" L. $5. *From the author's collection.*

Wood puzzle. 9" H x 13.5" L. 1970s. $15. *From the author's collection.*

Wood puzzle. 7.5" H x 12" W. $80.

Section 03.112 Religion and Elephants: Ganesha and Nativity

This subsection presents images of elephants used in conjunction with or as religious entities. For example, Ganesha (or Ganesh) is a Hindu god, a son of Shiva, an elephant-headed deity, who is believed to be capable of removing obstacles and bestowing temporal and spiritual accomplishment. In India, traditionally, no important undertaking is begun without first invoking Ganesha to ensure success. There are several Ganeshas in this category in different poses and materials, and some figurines, with other gods that ride atop elephants in their standard idolization configuration.

There are also several nativity elephants (as part of complete sets) from Lenox and Goebel/Hummel, and by individual artists such as Ferrandiz Juan and Prof. Kuolt Karl. Another elephant in the Hindu religion is Gajendra, who was lord of the elephants. According to legend, Gajendra entered a watering hole and was beset upon by a serpent. He raised a lotus flower to Vishnu and was rescued; see the small bronze of a Gajendra image in the Metal Figurines and Sculptures section.

Jumbo nativity elephant by Hummel/Goebel of Germany. Impressed on side, "SKROBEK," for Gerhard Skrobek, who is a master sculptor, joining Goebel in 1951. It also has "Trademark 6" impressed. Made for the Hummel 260 Nativity set. 11" H x 12.5" L. 1990s. $500. *Courtesy of Brenda McKenrick.* (E2D4).

All white bone china elephant from a limited edition Lenox Nativity collection. 8.75" H x 10" L. 1996. $50. *Courtesy of Joan.*

This is another version of the Lenox China Jewels Nativity elephant. This model is decorated with 24 kt. gold and china "jewels" in shades of blue. Discontinued in 2000. 6.5" H x 8.5" L. $150-200. *Courtesy of Carolyn.*

Bisque porcelain Chinese Immortal on Elephant; from a set of 16. Immortals are Hinayana Buddhist Arhats, the second highest stage of achievement. 12" H x 12" L. 1980s. $150-200. *From the author's collection.*

Hand-poured and finished Bronze Guanyin on adorned elephant. In China Guanyin is the popular name for Avalokitesvara: "He who perceives the cries of the world." He is one of the four Great Bodhisattvas recognized in China, presiding over the waters. He may be the most popular and venerated Buddha in China. Elephants are symbols of strength and prudence in Buddhism. This one is three-toed, which represents the Buddha, Dharma, and Sangha. The figure of Guanyin is clothed and ornamented as a prince and is seated in royal ease in the lila asana position. 19" H x 11" D x 25" L and weighs 47 lbs. 1990s. $900. *Courtesy of Artasia.*

Sandstone Ganesha. 8" H. 1990s. $50. *Courtesy of Abraham International.*

Hindu Bronze Ganesha Statue. 17" H (with umbrella) x 8" L. $100. *Courtesy of Silk Road Trading Concern.*

Marble Ganesha from Mandalay, Burma. 34" H. c. 1800. $4500-6500. *Courtesy of Abraham International.*

Carved Burmese jade Buddha sitting on an elephant, with a kid on his left and a bat on his head. The pronunciation of bat in Chinese is the same as for happiness. Chinese believe that bats can bring them happiness. Buddha is the most popular god in the East. 13" H x 11" W x 4" D. weighs 10 lbs. $1500.

Ganesha made from suar wood carving by master carver I Nyoman Subrata. The Hindu Ganesha, a half-man half-elephant demi-god is worshiped for the protection he provides and the wisdom he exudes. Ganesha is also called Lambhodara, meaning big belly, and Winayaka, the leader in heaven amongst the angels and gods. 13" H x 12.5" L. 2001 $175. *Courtesy of Novica.*

Section 03.113 Salt and Pepper Shakers

This subsection presents images of elephant salt and pepper shakers. Like elephant bookends, you could concentrate on this category and have a very satisfying collection.

Salt and pepper shakers are also chess pieces. Has "H83" on bottom. These are featured on the cover of a salt and pepper shaker book by *Tompkins and Thornburg 2000*. 3.75" H x 2.25" L. $40. *From the author's collection.*

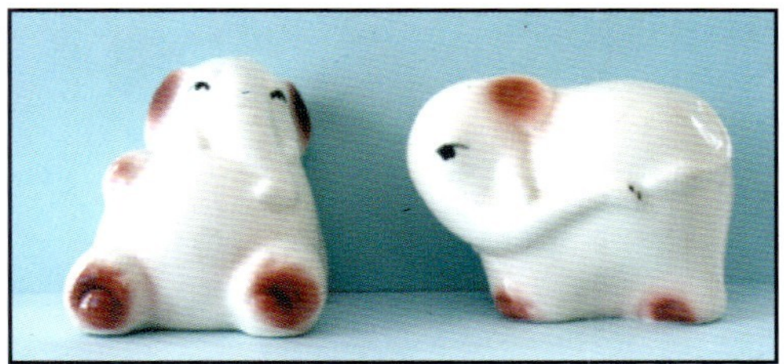

Ceramic salt and pepper shakers. 3" H x 2.5" L and 2.5" H x 3" L. $10. *From the author's collection.*

"Elephant and Boy" shakers are a joined pair of figures: a removable boy rider, which has 2 holes and the elephant named Sabu, with three holes. Ceramic Arts Studio. Elephant: 5" H x 3.75" L. Boy: 2.75" H x 1" L. 1952. $150-225. *Courtesy of Sue and John Horn.*

Ceramic salt and pepper shakers, marked "VCCTI." 4.75" H x 2.25" L. $7. *From the author's collection.*

Pair of salt and pepper shakers put out by the Ceramic Arts Studio, Madison, Wisconsin. These little boy and girl elephants are dressed alike in black dotted turquoise, shorts and a pompom hat for him, and a skirt for her with a bow on her head. Both have the black ink stamp mark on their bottoms. She is 3.25" H, he is 3.5" H. $55. *Courtesy of Sue and John Horn.*

Shakers from Japan. $7-10. *From the author's collection.*

Section 03.114 Salt Dips

This subsection presents images of salt dips or dishes.

Glass open salt with enameled elephant in the bottom. Made by Hoffman in Czechoslovakia; the Hoffman butterfly mark is in the lower left. 2" H x 2.75" L. 1920s. $40. *From the author's collection.*

Anodized brass salt dip. 1.25" H x 2.75" L. $10. *From the author's collection.*

Pottery salt dip with name "Pablo" followed by other script on bottom; probably from Mexico or Brazil. 3" H x 6" L. 1980s. $10. *From the author's collection.*

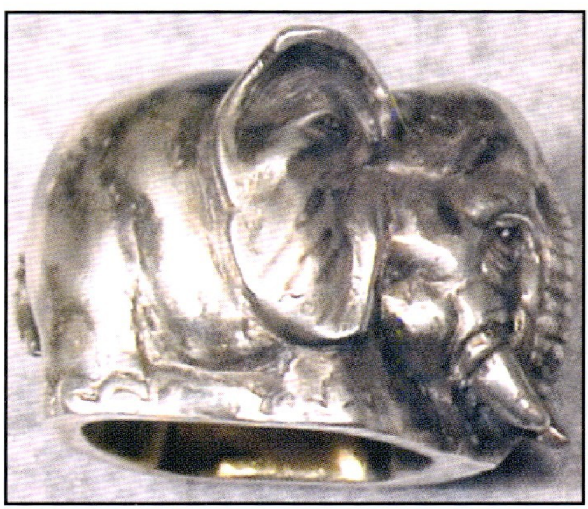

Silver salt dish from Faberge. c. 1900. $1000-1500. Faberge made several carved mineral elephants that featured precious stones as eyes.

Section 03.115 Scales and Weights

This subsection presents images of a scale and opium weights.

Bronze set of 6 east Asian opium weights, made in Thailand. These are newer representations of the opium weights made in Asia for hundreds of years. Older sets can range into the thousands of dollars. Value depends on: number of weights in the set, the sameness of the forms of the figures within a set, and the number of large sizes in a set, which is specified in ticals. c.1990s. $20-30. *From the author's collection.*

Bathroom scale with elephant image by Borg; fat elephant is saying: "Tomorrow I give up peanuts." 9" H x 12" L. c.1980s. $10. *From the author's collection.*

Section 03.116 Sewing

This subsection presents images of sewing-related paraphernalia.

Ceramic thimble by Lee, and Chinese resin thimble. 1.75" H x 1.25" L and 2" H x 1.5" L. Lee: 1984. Chinese: c.1990s. $8 each. *Courtesy of Adele Verkamp.*

Aunt Martha's iron-on transfers #3421 for tea towels. c.1950s. $20. *From the author's collection.*

Worcester Salt Company advertising needle case. 3" H x 1.75" L. c.1950s. $15-20. *From the author's collection.*

Sterling silver pincushion hallmarked with maker's mark "AandL" (probably Addie and Lufkin). Made in Birmingham, England. 1.25" H x 1.875" L. Hallmarked 1906. $100-150. *Courtesy of 20TH Century, Sydney, Australia.*

Ceramic pincushion by Ardalt. 4" H x 6" L. 1950s. $20. *From the author's collection.*

Silver-coated pot metal Victorian pincushion with blue felt cushion and original needles and pins. 2" H x 3.5" L. c.1890s. $75-100. *From the author's collection.*

Celluloid thimble holder; elephant is on a stand with a thimble holding piece next to him. Impressed "GERMANY" on the bottom, along with some pencil markings. 1.25" H x 2.5" L. Early 1900s. $50-75. *Courtesy of Chuck.*

Section 03.117 Shot Glasses

This subsection presents images of shot glasses.

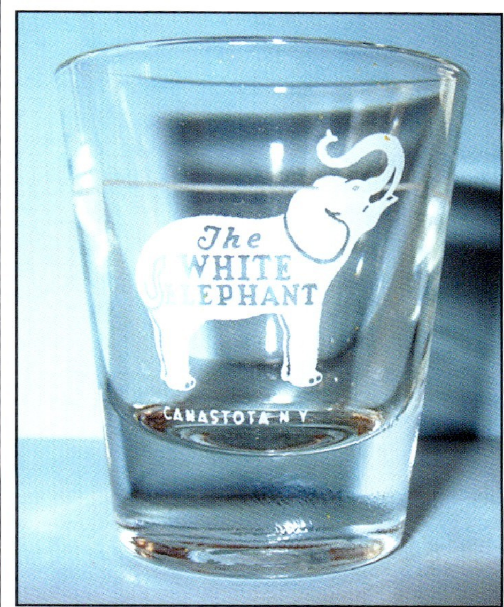

Clear shot glass with white detailing from "The White Elephant" in Canastota, New York. 2.375" H x 1.875" diameter at top. c.1940s-50s. $8. *From the author's collection.*

Clear shot glass with blue detailing; souvenir from Mammoth Mountain, California. 2.375" H x 1.875" diameter at top. c.1980s. $5. *From the author's collection.*

Section 03.118 Signs

This subsection presents images of signs with an elephant on them. Most such signs are advertising signs with the elephant as a logo or trademark of the company or entity, for example, Imported Elephant Beer or Carlsburg Elephant Malt Liquor. There are reproductions of older signs that cost much less than the ceramic-over-tin originals, which can run to several hundreds or even thousands of dollars.

Carlsburg Malt Liquor plastic bar sign in wood frame. 18" H x 15" L. 1980s. $10-15. *From the author's collection.*

Building sign from the Elephant and Castle Restaurant in Victoria, British Columbia. approx. 6 feet H. x 3 feet L. 2001. $500-1000. *Courtesy of Adele Verkamp.*

Double-sided die-cut porcelain sign advertising Bears cigarettes. 19" H x 7" L. 1930s. $1600+. *Courtesy of Nicholas Ciovica.*

Rare Eagle Cigarettes porcelain advertising sign manufactured by the British American Tobacco Co. Ltd. in Thailand. The old-styled Thai language at the top of the picture means "Don't forget to have one pack of EAGLE cigarettes in your pocket." There are also Chinese characters along the sides of the picture. 30 W" x 40" H. 1912 stamped on sign. $500+. *Courtesy of Jack.* (E2D4).

Plastic 3-dimensional sign advertising Imported Elephant Beer by the Anheuser-Busch Company; Item #530-020. 17" H x 18" L. 1987. $35. *From the author's collection.*

Tin sign advertising with mastodon or woolly mammoth for Rock Ice Amber Lager by the Latrobe Brewing Company in Latrobe, Pennsylvania. Approx. 22" diameter. 1994. $20. *From the author's collection.*

Lighted plastic sign box for Elephant Red beer of the Anheuser-Busch Co. 15" W x 15" H. 1990s. $35-50. *From the author's collection.*

Reproduction poolroom plastic sign advertising Brunswick-Balke-Collender Co. billiard and snooker tables. Item #E56 1990s. $100. The original was made c. 1923.

Die-cut enameled tin sign for Fremlins Ale. 12 H x 16" L. c.1940s-50s. $100. *Courtesy of John Stadnicki.*

Single-sided porcelain sign advertising for ESSO Elephant Kerosene. This rare sign is enameled and was issued by the Esso Oil and Kerosene Company (the Indian Esso company), which at that time was located in Bombay, India. This may be the only pre-war kerosene sign ever made by Esso The sign is created with a stencil technique showing high relief. It shows the classic Esso trademark with inscription: "ELEPHANT KEROSENE ." Has four screw holes. Signed "Bengal Enamel" (manufacturer's mark). 24" H x 12" L. early 1920s-30s. $350-400. *From the author's collection.*

Vintage enameled tin sign advertise the Platignum fountain and ballpoint pens. 12" W x 9.5" H. c.1950s. $50. *From the author's collection.*

Reproduction of porcelain-over-tin sign for Elephant brand petroleum from the Vacuum Oil Company; from one of Vacuum Oil's early foreign branches in India. This sign was made from an original watercolor rendition in the Mobil archives, and produced by Graphics Express. 1990s. $30. *Courtesy of Graphics Express.*

A Toppie electric sign featuring that infamous elephant that is the logo for Top Value Stamps. Front is rigid plastic and the box is metal. Label on side says "Hal MFG. Co. 3116 Spring Grove, Cincinnati 25, O. Kirby 1-2505. Sign and Pictorial Union Local 224, No. 38." Pullstring lights up rod; the light bulbs are inside. 15.25" H x 24.5" L. c.1950s. $300+.

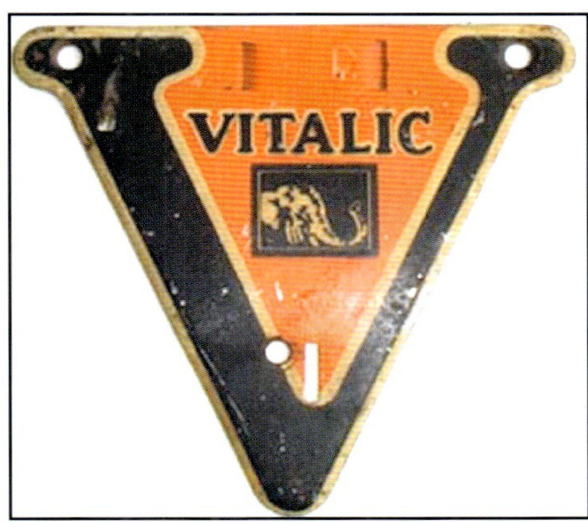

An enameled triangular tin sign advertising Vitalic products (see other Vitalic items in the Advertising Figurines and the Tins categories). 1940s-50s. $75-100.

Colored-foil-covered heavy cardboard die-cut Kraft Co. advertising sign of elephant's head with gold harness trim. 24" H x 33" L. c.1960s. $25-50. *From the author's collection.*

Section 03.119 Silverware and Utensils

This subsection presents images of silverware and other kitchen utensils. Arthur Court makes several items in this category include a great matched pair of salad utensils valued at about $30.

Flatware handle depicts three elephant heads on the handle, one facing front, one right, one left, a typical "Erawan" (3-headed elephant) design from southeast Asia. Part of a 160-piece set in nickel bronze These sets were made by factories including the Siam Bronze Factory in Bangkok, Siam. The myth has it that locals collected the ammunition casings after the Korean war sold them to the factories that melted them down to make these sets. c. 1950s. $150-200 set. *From the author's collection.*

A ceramic spoon rest from Bouzaki Gifts in Florida. 2.5" H x 7.5" L. c.1980s. $8. *From the author's collection.*

Section 03.120 Snuff and Opium Bottles

This subsection presents images of elephant snuff and opium bottles.

Reverse-painted crystal snuff bottle by artist Shao Jian Long, one of the top artists at the famous Xisan Academy of Inside Painting, and one of the finest in China when it comes to painting realistic animals. 3" H. 2000. $60. *Courtesy of The Snuff Bottle Club.*

Snuff bottle carved from agate. $50. *Courtesy of Manny Fellouzis.*

Chinese enameled metal snuff bottle with a coral-topped stopper. 3" H. $75-100. *Courtesy of John Rahmeyer.*

Chinese or Japanese bottle carved ivory and cold-painted. The man's head is the stopper. 2.75" H x 2" L. $250. *Courtesy of Donald Kieran Austen.*

Section 03.121 Soap and Soap Dishes

This subsection presents images of elephant soap and soap dishes.

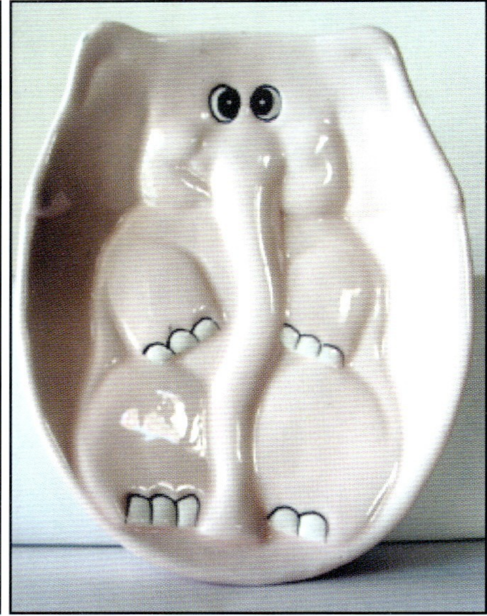

A ceramic pump-type liquid soap dispenser by Blokker. 7" H x 3.25" L. c.1980s. $7. *From the author's collection.*

Ceramic soap dish. 5" H x 4" L. c.1990s. $5. *Courtesy of Adele Verkamp.*

"Jungle Fun" soap and ceramic dish; made in China for Burlington Toiletries. 4.5" diameter. c.1980s. $5. *From the author's collection.*

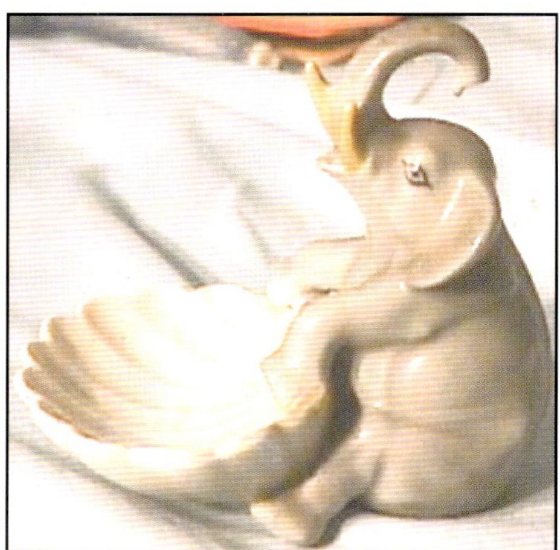

German Lustre ceramic soapdish. 4" H x 4" L. c.1930s. $25-35. *Courtesy of Larry's Antiques LLC.*

Brass soap dish. 3.25" H x 8.5" L. c.1980s. $15. *From the author's collection.*

Avon 2-ounce soap and die-cut box. 2.5" H x 2.5" L. 1985. $5. *From the author's collection.*

Section 03.122 Stamps, Covers, Cancels and Postmarks

This subsection presents images of elephant on stamps and postal stationary like envelopes. Stamp collecting (philately), with a special interest is called topical collecting. By combining marvels of miniature art and printing technology, stamps, with a love of elephants, you arrive at a result that is wonderful to collectors. Although there are none shown here, there are several stamps wherein the paper used has an elephant watermark. Many are listed in the expanded stamp list available from the American Topical Association; their address is in Section 4.

Ivory Coast stamp set of 3; *Scott Catalog #167-#169*. 1 Oct 1959. $1. *From the author's collection.*

A First Day Cover (known as an FDC); sometimes just a stamp and a first day postmark on an envelope; other FDCs combine the art of a stamp with, usually, a larger drawing by an artist (called a cachet) on an envelope. The envelope is then taken to a particular post office on the first day the stamp is issued and stamped with the day's postmark. This one features a *Scott Catalog #C138* stamp from Republic of Guinea and Commemorates the World Wildlife Fund. 1977. $3-5. *From the author's collection.*

A regular (non-cacheted) first day cover with a complete set of Laos elephant stamps *Scott Catalog #41-47*. Cover postmark dated 17 March 1958. $25-35. *From the author's collection.*

A ceramic stamp moistener made in Germany. A sponge placed in the center would have been used to moisten the glue on the stamp. 3" H x 1.75" L. 1905. *Courtesy of Jules.*

Thailand stamp showing Queen Suriyothai riding an elephant; *Scott Catalog #563* (part of a larger set). 25 Oct 1970. $1. *From the author's collection.*

Bulgaria stamp; elephant in the Sofia Zoo; *Scott Catalog #3329* (part of a larger set). 20 May 1988. $.15. *From the author's collection.*

A proof stamp sheet, from North Borneo, of *Scott Catalog #141*. 1909. $500. *Courtesy of Simon Andrews Stamps.*

A stamp from Monaco from a set titled: "Five Weeks in a Balloon," *Scott Catalog #340*. 7 June 1955. $0.25. *From the author's collection.*

Section 03.123 Stickers and Decals

This subsection presents images of elephant stickers and decals.

Decal for the Connie Mack Golden Jubilee 50 year commemorative, made for Connie's 50 years of service to the A's (Philadelphia Athletics) baseball team. The elephant has been the team's mascot, off and on, since Connie decided to use it in the early 1900s. 4.5" H x 4.5" L. 1950. $30-40. *From the author's collection.*

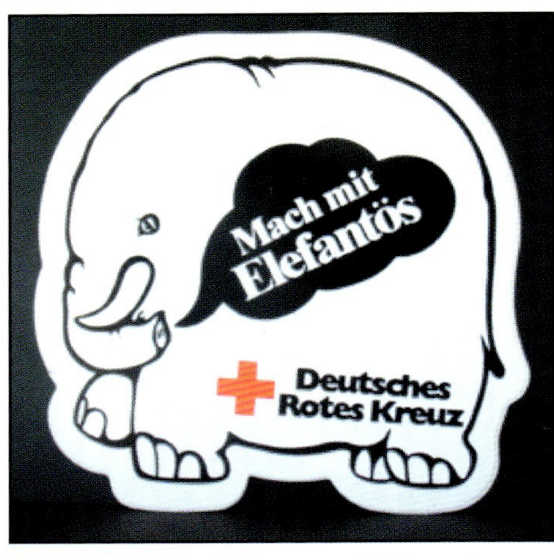

German Red Cross sticker. 4.5" H x 4" L. c.1990s. $5. *From the author's collection.*

Package of padded vinyl stickers from Hallmark Cards. c.1990s. $5. *From the author's collection.*

Section 03.124 String Holders

This subsection presents images of elephant string holders.

Porcelain string holder made in Japan. 5.5" H x 4" L. c.1980s. $10-15. *Courtesy of Adele Verkamp.*

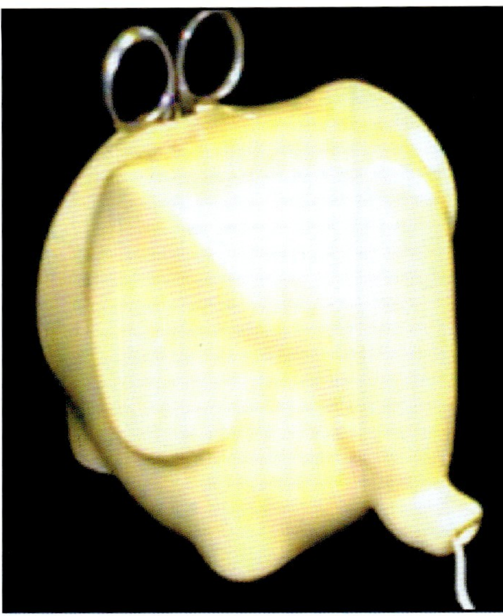

Porcelain string holder marked on bottom: "MADE IN ENGLAND." 5" H x 4.5" L. 1930s. $60. *Courtesy of Kenny Bogle.*

Section 03.125 Stuffed Elephants and Dolls

This subsection presents images of various stuffed elephants. No, not taxidermy-wise, but cute, little fluffy elephants stuffed with cotton or other fill! Steiff is one stuffed animal manufacturer that has produced many elephants. In fact, Margaret Steiff's first stuffed animal was a hand-made elephant that started the company in 1865. There are other Steiff elephants in the Toys and Games category. There are scores of Steiff stuffed elephants because the company has been making them for so long. I could do a book just on Steiff elephants, although other Steiff books have many of them (see: *Hockenberry*, 2000).

A recent stuffed-elephant phenomenon is the Beanie Baby by Ty Corporation. I included the Beanie Elephant Royal Blue Peanut. Other Beanies include: Trumpet, Righty (1996), Righty 2000, Horton (from the Dr. Suess stories), Peanut Light Blue, and Lefty. CVS makes a similar spotted "beanie" called Misfit and there are mini-"beanies" popping up all over, usually as co-branded versions to advertise a McDonalds or something.

Large Discovery Channel plush stuffed elephant. 10" H x 27" L. c.1990s. $35-45. *Courtesy of Ted Capell.*

This Beanie Baby, Peanut The Elephant, royal blue version, is rare as far as Beanies go. Style #4062 by Ty Inc. 1993. $300-2500 depending on date, tags and authentication. *Courtesy of Jane Benson.*

Elephant from Softball Creatures Series by J. Miller Clark. 3" H x 2.5" L. 1980. $10. *From the author's collection.*

Baki elephant made by German company, Baumann and Kienel-Baki. Baki was founded in 1946 in Flensburg /Schleswig-Holstein. The elephant is wool-filled with glass eyes. 14.2" H x 13.4" L. 1960. $50. *Courtesy of Stefan Eichhorn.*

Puffalump stuffed elephant from Fisher-Price. 10" H. 1993. $20. *Courtesy of Kristine Freeland.*

Ringling and Barnum, Bailey Brothers Circus stuffed elephant with blanket and plastic tusks. 9.5" H x 20" L. 1980. $25-30. *From the author's collection.*

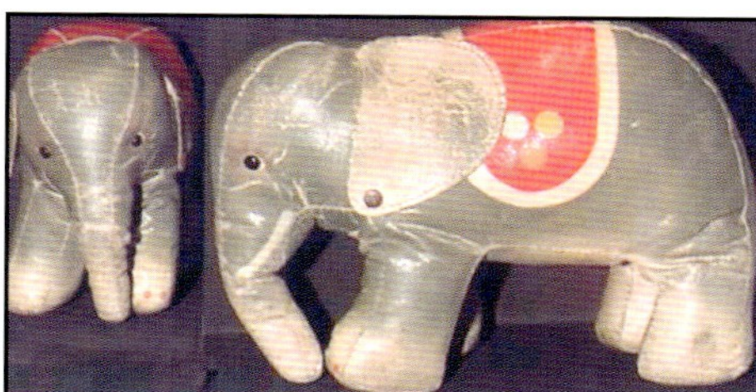

Unusual Steiff "bath-time" elephant with FF button, gray oilcloth, black glass eyes, and painted saddle. See this elephant in *Koskinen, p. 91*. 4.5" H x 7" L. c.1939. $280-425. *Courtesy of Fiona Miller of The Old Bear Company.*

Steiff stuffed elephant with red collar cloth and felt ears. c.1950s. $100.

Omar A. Pachyderminsky is the baby of the Boyds elephant family. He was issued in 1993 and retired in 1994. 7.5" tall. $35-40.

Section 03.126 Tape Dispensers

This subsection presents images of elephant tape dispensers.

Brass tape dispenser. 3.5" H x 4.5" L. c.1980s. $25. *Courtesy of Adele Verkamp.*

Ceramic tape dispenser with: "MJA 81" impressed on bottom. 4" H x 7.5" L. c.1980s. $15. *From the author's collection.*

Section 03.127 Teapots

This subsection presents images of elephant teapots. Teapots are a natural functional match for the elephant form, the trunk being a natural spout, the body a vessel and the tail a natural handle.

Ceramic 2-headed highly glazed teapot made by Wood's of England. 9" H x 11" L. c.1970s. $50. *From the author's collection.*

Ceramic teapot made by Fitz and Floyd. 9" H x 7.5" L. c. 1992. $35. *Courtesy of Adele Verkamp*.

Porcelain enamel-over-metal teapot; trunk is removable handle. Made for Kamenstein in Taiwan. 11" H x 9.5" L. c.1970s. $50. *From the author's collection*.

Shawnee yellow lidded teapot, with imprinted mark: "Patented USA." 7.5" H x 7" L. c. 1940s. $175-250. *Courtesy of Christine Vattuone*.

Petite ceramic teapot made in China. 6 " H x 5" L. c.1970s. $25. *From the author's collection*.

Ceramic hand-painted teapot made in Japan with bamboo handle and rider on cover. 6" H x 8" L. c.1960s. $30-40. *From the author's collection*.

Shawnee elephant teapot with typical marking: "USA" impressed on bottom. 7.5" H x 7" L. c. 1940s. $125-175.

Porcelain teapot marked: "Andrea by Sadek" from Thailand. 7" H x 9" L. $30-40. *Courtesy of Adele Verkamp*.

This appears to be an antique brass, elaborately caparisoned teapot or pitcher of some sort. There is a round opening in the elephant's back covered by a finial that is also an elephant. The trunk is a pouring spout. But it is very large and very heavy, weighing almost 15 pounds with no obvious handle so a tea server would have to be strong to pour the tea. 10" H x 12.25" L. early 1900s. $200. *From the author's collection*.

Section 03.128 Tiles

This subsection presents images of tiles with an elephant image on them. Delft tiles, especially from the 1700s and before, are most sought after.

Round painted plaster tile. 6.25" diameter. c.1990s. $5. *Courtesy of Adele Verkamp.*

Ceramic tile from Dedham with usual crackle glaze. 5" H x 5" L. early 1900s. $600-800. *Courtesy of Leonard Williams.*

Ceramic tile from Delft area of Holland; specifically a northern province called Friesland. Polychrome coloring/glazing is unusual. Other earlier Delft tiles have a blue elephant and date back to the 17th century. 5" H x 5" L. 1800s. $100. *From the author's collection.*

Section 03.129 Tins and Canisters

This subsection presents images of tins and canisters that held various substances over the years: from tea, tobacco, salt and nuts, to rubber soling compounds. Other tins I am familiar with include: Bazzini Nuts of New York (a relatively rare tin worth in the $300-500 range), Burch's Popcorn, Shedd's Peanut Butter, and many different Lipton and Tetley Tea tins.

Embossed figural elephant Earl Grey Tea tin by Williams and Magor, London. 6.625" H x 4.25" diameter. c.1990s. $15-25. *From the author's collection.*

Colorful graphics on this Lipton Tea tin made in Canada. 5.375" H x 3". c.1940s. $10-15. *From the author's collection.*

Cardboard Ivory Salt container produced by the Morton Salt Company. 1-pound size. c.1940s-50s. $5-10. *Courtesy of Barb Cramer.*

Manley Jumbo elephant popcorn tin; graphics are similar to those used on a Burch popcorn tin. 9.5" H. c. 1940s. $20. *Courtesy of Mary DuQue.*

Vitalic brand rubber tube repair kit tin manufactured by the Continental Rubber Works, Erie, Pennsylvania. 4.5" H x 2" diameter. c.1940s-50s. $25. *From the author's collection.*

A Republic of Green Tea Company traveler's tea tin for "Republic Green Chai: Merchant's Tea" from Novota, California. 5.5" H x 3" diameter. 1999. $5. *From the author's collection.*

A marjoram spice tin with the Crown Colony elephant logo. 2.625" H and 2.5" L. 1960s. $3. *From the author's collection.*

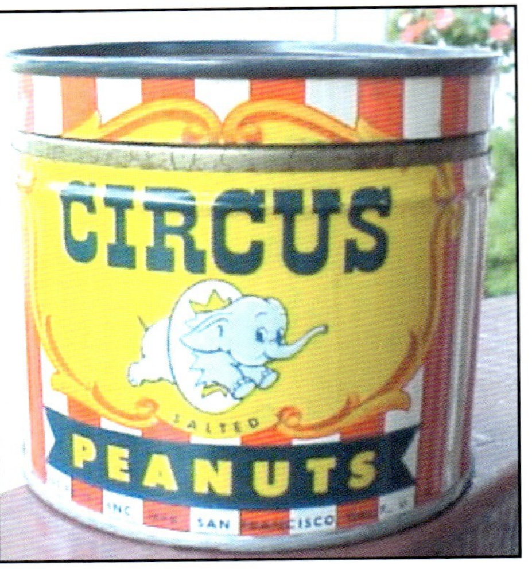

A Circus brand blanched Virginia peanuts tin with elephant jumping through a circus hoop by the Circus Food Company of San Francisco, California. 8 oz. 3" H x 3.375" diameter. 1946. $50-75.

Jumbo Elephant Soling Compound tin made in Australia. 1.75" H x 2.75" diameter. $10. *From the author's collection.*

New World 50-gram tobacco tin by Elephant and Castle. Produced by McConnel of England (now in Germany) and imported to the U.S. by Marble Arch. Tins exist for Cromwell tobacco and Stout tobacco from Elephant and Castle, with the same graphics in brown and deep pink, respectively. 1991. $40-50. *Courtesy of John H. Eells.*

Lloyds "Bondman" Tobacco tin from Richard Lloyd and Sons, London with trademarked impressed elephant. 1.75" H x 2.625" diameter. $10-15. *From the author's collection.*

An Elephant brand salted peanuts 1/2 lb. tin from the Superior Peanut Company of Cleveland, Ohio. 4" H x 4" L 1906. $350-600. *Courtesy of David.*

Sharpes Toffee tin has a photo-image of a elephant on its hinged lid. It is press stamped "Made in England By Edward Sharpes and Sons Ltd of Maidstone Kent". 6" W x 4" L. c.1930s. $10. *Courtesy of Val Leeds.*

Two royal Indian elephants are in the center scene on this Celestial Seasonings Tea tin. 6.125" H x 8.125" L. 1989. $5. *From the author's collection.*

Bears medium flake cut tobacco tin with trademark elephant image; made by the Imperial Tobacco Co. of New Zealand Ltd in Wellington. 2.25" H x 3.25" diameter. $10-15. *From the author's collection.*

Ten-pound Elephant salted peanuts tin by the Superior Peanut Company of Cleveland Ohio. 11.25" H x 7.5" diameter. $250-350. *From the author's collection.*

Oilcan produced for Heavy Duty Oil. 1-quart size (there is a 2-gallon rectangular can as well). c. 1950s. $175-225. (E2D4).

A tea tin (possibly from Ceylon Tea) for Maravilla tea (an herbal tea made from the root of the Maravilla high desert plant). $50.

Section 03.130 Tobacco Silks, Felts, and Tags

This subsection presents images of tobacco-related paraphernalia. Tobacco silks and felts were placed in tobacco pouches as a premium - like a Cracker Jack prize. Felts and silks are often sewn into quilts! According to Lou Storino (1995) tobacco tags were used to mark or brand tobacco plugs, because cheaper brands were often substituted in more expensive boxes. Lorillard and Pioneer Tobacco Cos. started the practice using wood-tagged plugs. Tin tags were first used by Lorillard. A tag was initially applied to outer leaf but hidden. Then Ben Finzer Tobacco Co. applied them to the outside so a user would see it and not chomp down on the tag, possibly breaking teeth!

Tobacco felt with a Siam (now Thailand) flag. 8" H x 6" L. c.1910-1916. $15. *From the author's collection.*

Art Nouveau style tobacco silk with the crest of Ceylon. Manufacturers began substituting silks for insert cards beginning about 1909. 3.25" H x 2.25" L. c.1900. $20. *From the author's collection.*

Tobacco silk issued in pack of Egyptienne Luxury cigarettes. The design is from a series called Orders and Military Medals and this one is "Order of The Elephant" which depicts a military medal of that name. 3.875" H x 1.875" L. c.1910. $20. *From the author's collection.*

Tobacco silk with a Siam elephant, Erawan three-headed elephant image and crossed swords. From Canadian Misc Series SC12. 3" H x 1.75" L. c.1910. $10. *From the author's collection.*

Wooden tobacco plug from Red Elephant tobacco made by Brown Brothers. 1" diameter. c.1870-1930. $15-25. *Courtesy of David Wampler.*

Section 03.131 Toothpick Holders

This subsection presents images of elephant toothpick holders. Another toothpick holder I am aware of is a brass one by Florenza.

Ceramic toothpick holder made in Japan, possibly by Enesco. 3" H x 2" L. 1990s. $5. *From the author's collection.*

Slag glass toothpick holder by Degenhart (before the "Boyd" era) with the D in heart logo (removed in 1978). This was the name for the Crystal Art Glass Company in Cambridge, Ohio. 2.5" H x 4" L. c. 1975. $25. *Courtesy of DeeJay's Glass.*

Ceramic toothpick holder doubling as an hors d'oeuvre holder by Starnes of California. $20-30.

Section 03.132 Toys and Games

This subsection presents images of recent, vintage and some antique toys. Many other elephants in this encyclopedia could be thought of as toys, such as the plastic figurines or stuffed elephants. But I think the ones in this category were specifically made to be toys to be played with, and not ones I could place in the other categories.

Large carved solid wood rocker with stained saddle and cap. 20" H x 24" L. 1990s. $200. *From the author's collection.*

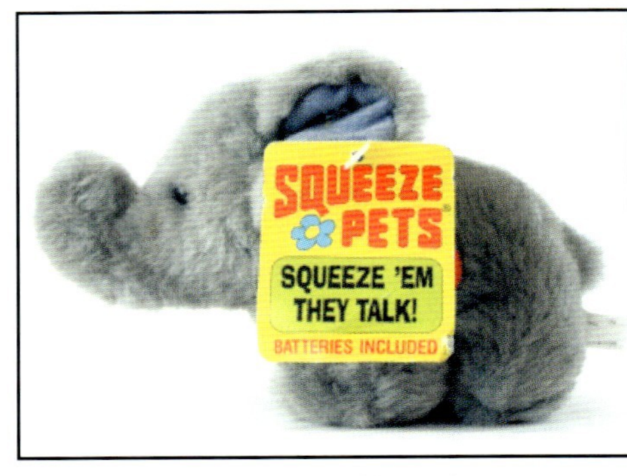

Fuzzy Squeeze Pet from China that roars when squeezed; item #SP01. 4.5" H x 8" L. c. 1992. $10. *From the author's collection.*

When turned on, this mechanical stuffed elephant walks and its trunk goes up and down; battery powered. 9" H x 15" L. 1990s. $25. *From the author's collection.*

Battery-operated walker with tethered remote control; has a trademark triangle with a T and M inside for Modern Toys of Japan. This is a similar to a toy called the Lambo Elephant that also has a magnetic trunk and metal logs and pulls a small wagon. 8" H x 6" L. 1940s-50s. $50. *From the author's collection.*

Tibetan hand-carved elephant wood marionette. Each joint is movable, including the trunk of the elephant. It is hand-painted and tied. The tail is made from hair. There are several varieties of these wooden marionettes made all over the world including the Far East and Mexico. 8" H x 10" L. 1920s. $50-75. *From the author's collection.*

Painted cast iron pull toy and bank. 2" x 3" L. 1920. $30-40. *Courtesy of christysgifts.com.*

A plastic Cybertron Transformer toy; item #DA5. This model is known as "Longhorn" and is made by the Takara Toy Co. Ltd. Japan. It transforms into a humanoid robot by itself and also joins two other robots to form a giant master robot known as "Magnaboss." 3" H x 5" L. 1996. $10-20. *Courtesy of Sean M. Kotran.*

Early Gong Bell pull toy of Jumbo the Elephant. One side is marked "The Gong Bell Mfg. Co. Made in USA" the other marked "copyright 1938 H. D. Allen." 8.5" H x 9' L. 1938. $150-200.

This toy is called Henry on the White Elephant; it was made by George Borgfeldt. It has a wind-up mechanism with key. 4" H x 8" L. c.1930s. $1400 with original box. *Courtesy of Harvey Kletsroc.*

Painted, cast iron pull toy and penny bank from Meir of Germany. 2" H x 3" L. 1920s. $300-450. *Courtesy of Tom Slabbinck.*

Elephant No. 2 puppet by Bob Pelham of England, who made some of the world's greatest puppets and marionettes from 1947 through the 1980s. 12" H. Similar models issued in 1963 and 1970. $150+. *Courtesy of Kim Ker.*

Schoenhut articulated wood elephant with original gray paint, glass eyes, leather ears, and rope tail. Its tusks are leather, painted white. 6" H x 9" L. c. early 1900s. $300. *Courtesy of Brad Maxwell.*

A teak game that entails each competitor placing a wooden plug onto the growing pile of plugs on the elephant's back until it topples. 8" H x 6" L. 1980s. $25. *From the author's collection.*

Recent Steiff Golden Age of the Circus Elephant and Calliope. It is the first in a series of five circus wagons, all numbered, limited editions. The Elephant on Wheels is a reproduction from Steiff's archive. The materials used and the production of the elephant follow that of the original. The elephant has a button in the ear and is made of mohair. The calliope is battery operated and features a music box. A Certificate of Authenticity is included. 13" H x 24" L. 1990s. $300-350. *Courtesy of Vivian.*

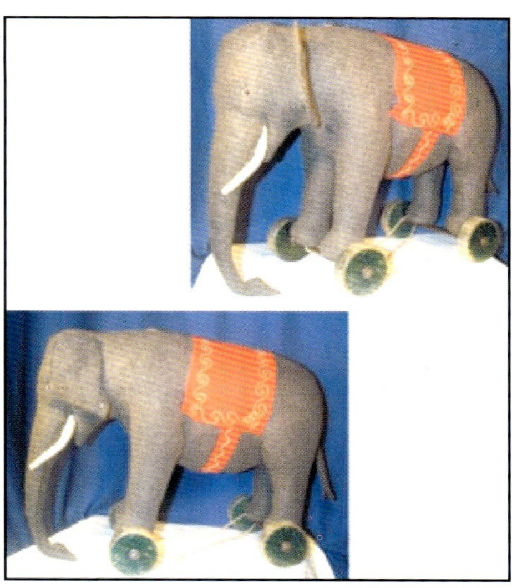

Rare vintage Steiff riding elephant with Wiegel ufel (a wheeled carriage). It has a felt-covered surface and metal wheels with a rubber surface. It also has a voice that emanates when a handle is turned. 19" H x 24" L. 1905-1915. $1500-2000. *Courtesy of Gabi Lutz.* (E2D4).

Wind-up tin elephant on bike with swirling ball. It comes boxed in the Schylling Collector Series Box. 11" H x 5.75" L. 1990s. $10. *From the author's collection.*

A tin walking elephant with bright lithographic paint; built in the US Zone of Germany. 4" H. 1940s-1950s. $150-200. *Courtesy of Paul Melito.*

Victorian trade card advertising Lavine Washing Powder. 2.5" H x 4.25" L. 1880-1910. $10-15. *From the author's collection.*

Victorian trade card advertising Bush and Bull Dry Goods in Watertown, New York. 2.5" H x 4.25" L. 1880-1910. $10-15. *From the author's collection.*

Section 03.133 Trade Cards

This subsection presents images of mostly Victorian era trade cards which were used to advertise products, much like today's business cards advertise a company and person. Many of the individual cards presented here are part of a set or series, some with all elephants, others with mixed subjects. Considering their age, many are extremely colorful and have finely detailed lithography. Tobacco cards were typically inserted into different brands of cigarettes to encourage product loyalty and increase market share.

A trio of Victorian trade cards included as premiums in packages of chocolate by the German firm Gerb. Stollwerck (Koln, Germany). They are numbers: 460-1, -2, and -3 in a series. 3.7" H x 1.9" L. 1908. $25 set. *From the author's collection.*

"Jumbo Must Go" trade card advertising Willimantic Thread. 3.375" H x 5.75" L. 1881. $10-15. *From the author's collection.*

Victorian trade card showing elephants working at a Liebig Meat Processing Plant (of Germany) in South America; part of a 6-card set. 2.75" H x 4.5" L. 1880-1910. $10-15. *From the author's collection.*

Victorian trade card advertising Fairbanks Standard Scales. This is Barnum's white elephant, called Toung Taloung (see Die-cuts category), being weighed. 3.125" H x 5.125" L. 1880-1910. $25. *From the author's collection.*

"Jumbo At The Opera" trade card by Clark's O.N.T. Spool Cotton thread. 3" H x 4.75" L. 1890s-1930s. $10-15. *From the author's collection.*

Two elephants keeping warm on a trade card advertising Adams and Westlake Oil Stoves. 3.5" H x 5.5" L. 1870s-1920s. $10-15. *From the author's collection.*

This trade card actually advertises elephants! Advertising is on the front and the back for Adam Forpaugh, Jr. and his featured show of 30 trained elephants. This includes Bolivar, the largest elephant alive (although Barnum's Jumbo was also alive at the time), and a boxing elephant. 4.25" H x 6" L. 1880s-1910. $45-65.

Trade card showing an "Asiatic Elephant" prepared as a premium for Dwight's Soda, issued by John Dwight. Number 34 in a set of 60. 2.5" H x 1.5" L. 1870s-1920s. $10-15. *From the author's collection.*

Trade card advertising Colburns Mustard. 2.75" H x 4.25" L. 1870s-1920s. $10-15. *From the author's collection.*

Victorian trade card advertising Ivorine Washing Powder. 3.125" H x 5" L. 1880-1910. $5. *From the author's collection.*

These are the connected tete beche inside panels of a 2-panel folder for Kerr's Six Cord Spool Cotton thread. Lithography by George Schlegel and Son. 3.125" H x 5" L. 1880-1910. $50. *From the author's collection.*

Victorian tobacco trade card given as a premium by Allen and Ginter Tobacco Company of Richmond, Virginia. It is #12, "Order of the Elephant, Denmark" out of a "World's Decorations" series of 30. Lindner, Eddy and Clauss, New York Lithographers. 2.75" H x 1.5" L. 1890. $10-15. *From the author's collection.*

Victorian trade card advertising Hartford Sewing Machines. 3" H x 5" L. 1880-1910. $15. *From the author's collection.*

Front panel of a rare 2-panel folder for Kerr's Six Cord Spool Cotton thread. Cards are tete beche. Also see next image. Lithography by George Schlegel and Son. 3.125" H x 5" L. 1880-1910. $50. *From the author's collection.*

Section 03.134 Trademarks and Logos

This subsection presents images of elephants that serve as either trademarks or logos of some sort for a particular company or product. There are other items in other categories that could have been placed here, depending on how I categorized. For example, an item can be both a tobacco tin, and have an elephant image as a trademark on it, but I placed it in the Tins and Canisters category. But if I did the reverse and placed all elephants used as logos or trademarks here, this category would be very large! Also see the Beer categories for items using the elephant as a logo or brand identifier.

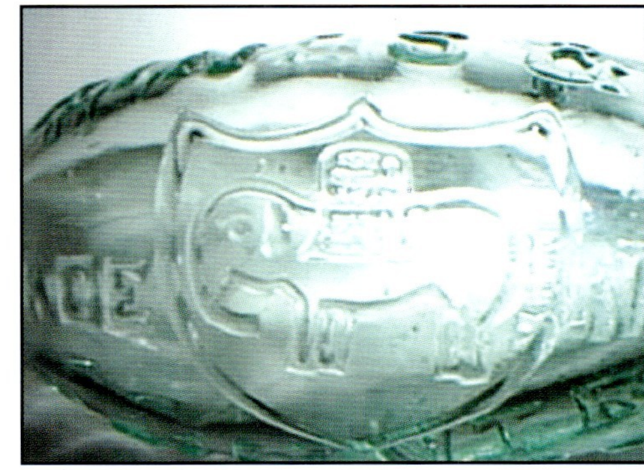

Trademark impressed elephant and howdah in shield on a Hinds and Co. bottle. 1880.

Image of the trademark winged elephant in a circle, for publisher, The Overlook Press of Woodstock, New York;. This particular image taken from my copy of *Travels On My Elephant* by Mark Shand. The logo was designed by Milton Glaser and is about .5" round. 1992. *From the author's collection.*

Trademark of Elephant brand Black Powder on a patch. Firm in existence since 1866. 2.5" H x 3.25" L. 1990s. *Courtesy of Joseph J. Borgatti Sr.*

Elephant trademark hallmarked on the bottom of a Reed and Barton 1906 pattern, "Francis I," bowl, pot, or spittoon. 1906. *Courtesy of Ron King.*

Elephant in square trademark printed on the bottom of Steubenville "Ivory" plate. c. 1950s-60s.

Section 03.135 Trivets

This subsection presents images of trivets which have a shape or image of an elephant. Trivets common today are often metal or ceramic versions of hotpads designed to place hot dishes upon. Earlier trivets had three, longer legs, and sat tripod-like, holding a pot or bowl, in a fireplace.

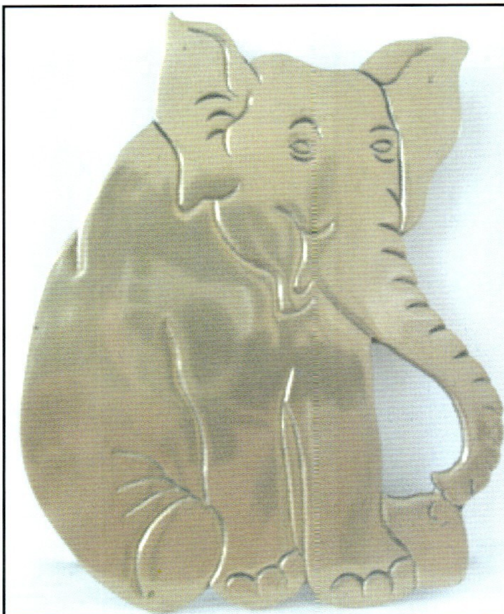

Brass trivet by Penco of Bradford, Massachusetts. 12" H x 10" L. c.1990s. $20. *From the author's collection.*

Aluminum trivet by Arthur Court, Massachusetts. 7.5" diameter. c.1995. $20-30. *Courtesy of Adele Verkamp.*

Section 03.136 Vases and Amphora

This subsection presents images of elephant vases. Some planters can be considered vases and vise versa, but if the market is calling it a vase - it is here.

Stoneware celadon bud vase with crazed glazing. 5" H x 4" L. 1950s. $20. *From the author's collection.*

Kay Finch pink elephant vase with an embossed mark: "Kay Finch California" in print, block type letters. 5.75" H x 5.5" L. c. 1940s. $125-150. *Courtesy of Penn Cove Antique Mall in Coupeville, Washington.* (E2D4).

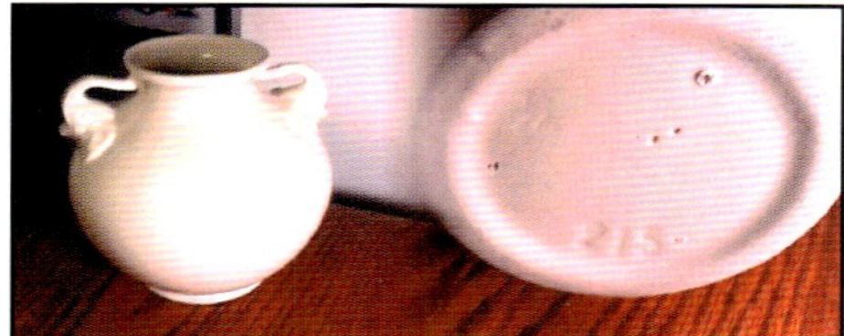

Rumrill pottery vase with elephant head handles. 8" H. c. 1930s-40s. $150-200.

namel-over-brass bud ase made in China. 3.5" I x 1.5" L. c.1950s. $10. *rom the author's ollection.*

Ceramic vase with "QQ" markings on bottom. 7" H x 4.5" L. c.1980s. $25. *Courtesy of Adele Verkamp.*

Carved wood vase. 8" H x 5" L. 1998. $50. *Courtesy of Janet Wojciechowski.*

Ceramic bud vase with "O125" impressed on bottom. 6.25" H x 3.25" L. c.1980s. $15. *From the author's collection.*

Ceramic celadon bud vase made in China for Baum Brothers. 11" H. 1990s. $20. *Courtesy of Tamara Fogg.*

Rare Victorian Staffordshire elephant spill vase has a standing elephant with nearside front and back legs free. The spill vase in the form of a tree rises behind the elephants back. Similar elephant illustrated on *Kenny 1998, 128*. 6.25" H x 6" L. c.1840s. $400+. *Courtesy of David Tulk of Madelena Antiques*. (E2D4).

Rare Art Deco pressed glass vase with elephant image designed by Pierre D'Avesn, pupil and sculptor of Rene Lalique. (D'Avesn created the famous Lalique Snake). This vase was created before D'Avesn entered the Daum studio. 9" H x 7" L. 1926-1930. $2000-3000.

Section 03.137 Wall Art

This subsection presents images of elephants that can be hung on a wall, but yet are not Paintings and Prints.

Copper sheet with relief elephant and foliage. 12" H x 9.5" L. 1990s. $10-15. *From the author's collection.*

CopperCraft framed copper etching by artist Joel Kirk. 7.5" H x 9.75" L. $50. *From the author's collection.*

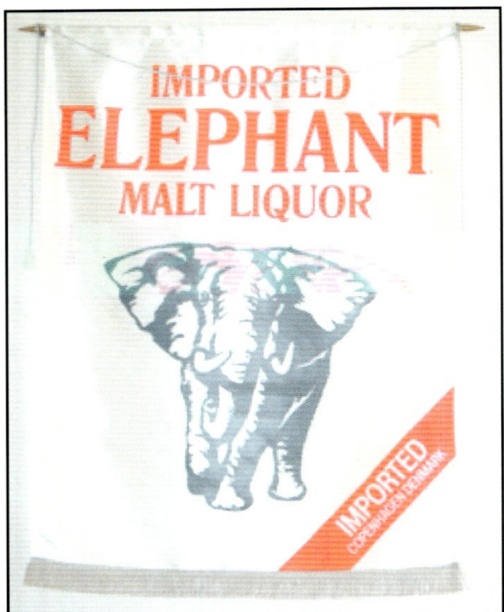

Elephant Malt Liquor silk bar banner with wood rod. 42" H x 30" L. 1980s. $10. *From the author's collection.*

Metal sculpture from Artesian House by artist O. C. Jere. 29" H x 31" L. 1995. $300. *From the author's collection.*

A large woven Renaissance-style tapestry inspired by the Imperial Elephant of India. Low warp weave from France with bold colors. The original of this antique reproduction is very large and is at the Louvre museum in Paris, France. The original was woven in Aubusson in the middle of the 19th century. 51" H x 68" L. 1990s. $600. *From the author's collection.*

Section 03.138 Wall Pockets

This subsection presents images of wall pockets.

Wallpocket marked California Cleminsons. He is decorated in the typical Cleminson's fashion of bright pastels and hand decorating. 2.75 Deep x 5.75" L. c.1950s. $60-75. *Courtesy of Mike Schneider.*

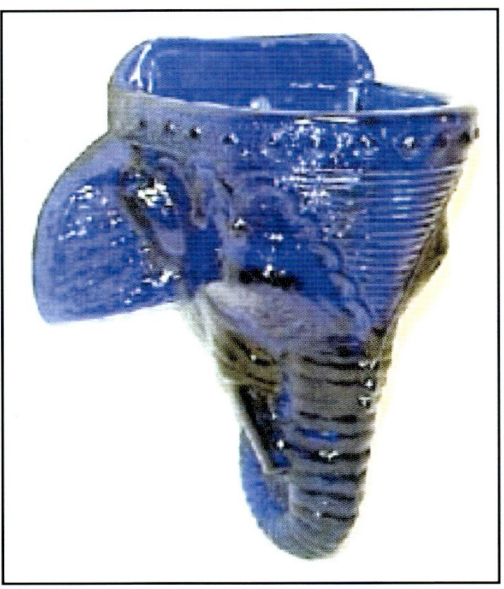

Cobalt glass wallpocket. 5" H x 4.5" L. 1990s. $10-15. Older original versions of these recent wallpockets, dating from teh Depression era, can be identified by their tapered hanging hale. *Courtesy of Elegant-Galleria.*

Section 03.140 Watering Cans and Sprinkler Bottles

This subsection presents images of elephant watering cans or bottles. One sought after item is a vintage laundry bottle.

Small aluminum watering can with: "Korea TT" impressed on bottom. 5.5" H x 8.5" L. 1994. $5. *From the author's collection.*

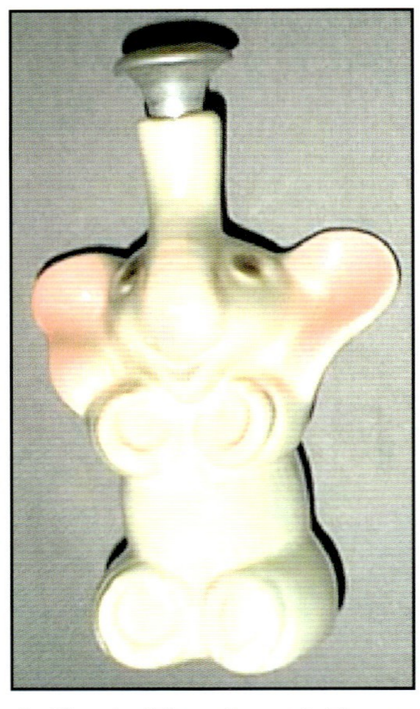

Ceramic "Jumbo" laundry sprinkler can with cork; one can be seen in (Bercovici, et al. 1998, 108). 7.5" H. c.1950s. $30-50. *Courtesy of Mary Beth.*

Section 03.139 Wallpaper

This subsection presents images of wallpaper.

Prepasted Border wallpaper by York. Design AY 6755-B. 5 yard spool x 10.5" H. 1990s. $20. *From the author's collection.*

Vintage wallpaper. c.1960s. $20. *Courtesy of Brian of Bentley's Sales.*

References and Resources

Many resources were used to generate information in this book; talking to antique and collectible dealers, auction realizations, my own experience, individuals supplying images, on-line web sites and various reference books. These references were used to identify items and find out about manufacturers and artists and establish or validate values. Most, if not all of the books listed actually have elephant collectibles in them; those that do not were used to gather manufacturer, category or artist information.

Most of the individuals contributing information about an item are listed in the "Courtesy Of:" part of each caption, if they so chose to be recognized. Many of them provide web sites or other establishments that you can visit.

In addition, I have compiled lists of additional elephant items in the following categories: Books, Coins, and Stamps, that are too exhaustive (many pages each) to include in this book. I can provide these if you contact me at my email address: conscioussystems@mindspring.com or mail to: Michael Knapik 15804 East Greystone Drive, Fountain Hills, AZ, 85268

References

Books

Aikins, Larry and Pat. *The World of Kreiss Ceramics*. L-W Book Sales, 1999.

Altman, *Book Of Buffalo Pottery*. Atglen, Pennsylvania: Schiffer Publishing Ltd., 1997.

Arbittier, Elizabeth & Douglas, Janet & John Morphy. *Collecting Figural Tape Measures*. Atglen, Pennsylvania: Schiffer Publishing Ltd., 1995.

Armke, Ken. *Hummel: An Illustrated Handbook and Price Guide*. Wallace-Homestead Book Co. 1995.

Atkinson Young, Nancy. *Breyer Horses, Riders, & Animals Molds & Models*. Atglen, Pennsylvania: Schiffer Publishing Ltd., 1999.

Barta, Dale and Diane, and Helen M. Rose. *Czechoslovakian Glass & Collectibles: Book II: Identification & Value Guide*. Collector Books, 1996.

Bercovici, Ellen. et al. *Kitchen, Bath, and Beyond*. Antique Trader, 1998.

Bess, Phyllis and Tom. *Frankoma: and Other Oklahoma Potteries*. Atglen, Pennsylvania: Schiffer Publishing Ltd., 2000.

Blake, Brenda. *Egg Cups, Illustrated History & Price Guide*. Antique Publications, 1996.

Bockol, Leslie. *Victorian Majolica*. Atglen, Pennsylvania: Schiffer Publishing Ltd., 1996.

Bredehoft, Neila. *Collector's Encyclopedia of Heisey Glass 1925-1938/With Price Guide*. Collector Books, 1986.

Breen, Walter. *Walter Breen's Complete Encyclopedia of U.S. and Colonial Coins*. New York, New York. E.C.I Press, 1988.

Browell, Felicia. *Breyer Animal: Collector's Guide*. Collector Books, 1997.

Bruner, Michael. *Encyclopedia of Porcelain Enamel Advertising*. Atglen, Pennsylvania: Schiffer Publishing Ltd., 1999.

Callow, Diana and John, and Marilyn and Peter Sweet. *Beswick Animals, 4th Edition*. The Charlton Standard Catalogue, The Charlton Press, 1999.

Carey, Susan S. and Ryan M, and Tara L. *The Beanie Encyclopedia: A Complete Unofficial Guide to Collecting Beanie Babies*. Collector Books, 1998.

Cast, Anthony and John Edwards. *Royal Worcester Figurines, 2nd edition*. The Charlton Standard Catalogue. The Charlton Press, 2000.

Chipman, Jack. *Collector's Encyclopedia of California Pottery*. Paducah, Kentucky: Collector Books, 1992.

Congdon-Martin, Douglas. *Figurative Cast Iron: A Collector's Guide*. Atglen, Pennsylvania: Schiffer Publishing Ltd., 1994.

Corson, Carol. *Schoenhut Dolls - A Collector's Encyclopedia*. Hobby House Press, 1993.

Curran, Pamela Duvall. *Shawnee Pottery, The Full Encyclopedia*. Atglen, Pennsylvania: Schiffer Publishing Ltd., 1995.

Dale, Jean. *Animals, Royal Doulton. 2nd Edition*. The Charlton Standard Catalogue. The Charlton Press, 1998.

Deel, Kathleen. *Napco*. Atglen, Pennsylvania: Schiffer Publishing Ltd., 1999.

Delozier, Loretta. *Collector's Encyclopedia of Lefton China Book II*. Collector Books, 1997.

Delozier, Loretta. *Lefton China: Price Guide*. Collector Books, 1999.

Dilley, David D. *Haeger potteries Through the Years*. Collector Books, 1997.

Divone, Judene. *Chocolate Molds: A History and Encyclopedia*. Oakton Hills Publications, 1987.

Dodge, Fred. *Antique Tins Identification & Values: Identification & Values, Books I and II*. Collector Books, 1994, 1998.

Dollen, Brenda and R. L. Dollen. *Collectors Encyclopedia of Red Wing Art Pottery: Identification & Values*. Collector Books, 2000.

Duke, Harvey. *Stangl Pottery*.

Ellis, Anita J. *Rookwood Pottery: The Glaze Lines*. Atglen, Pennsylvania: Schiffer Publishing Ltd., 1995.

Florence, Gene. *The Collector's Encyclopedia of Occupied Japan Collectibles: 5th Series*. Collector Books, 1990.

Forrest, Michael. *Art Bronzes*. Atglen, Pennsylvania: Schiffer Publishing Ltd., 1988.

Frick and Hodge. *Disneyana Collectors Guide to California Pottery*.

Garmon, Lee and Dick Spencer. *Glass Animals of the Depression Era*. Collector Books, 1993.

Gaston, Mary Frank. *The Collector's Encyclopedia of Limoges Porcelain, 2nd Edition*. Collector Books, 1998.

Giacomini, Mary Jane. *American Bisque: Collector's Guide with Prices*. Atglen, Pennsylvania: Schiffer Publishing Ltd., 1994.

Gibbs, Carl Jr. *Collector's Encyclopedia of Metlox Potteries: Identification and Values*. Collector Books, 1995.

Godden, Geoffrey. *British Pottery - An Illustrated Guide*. Barrie and Jenkins, 1975.

Guarnaccia, Helene. *Salt and Pepper Shakers; Identification and Values, vol. 1 to 4*. Collector Books, 1991.

Gwathmey, Emily, *An Enchantment of Elephants*, Clarkson N. Potter, Inc., 1993.

Hall, Doris & Burdell. *Morton Potteries*. L-W Publishing & Book Sales, 1995.

Hanson, Bob, Craig Nissen and Margaret Hanson. *McCoy Pottery: Collector's Reference & Value Guide Featuring the Top 100 Findables* (*McCoy Pottery Collector's Reference and Value Guide, Vol 2.*) Collector Books, 1999.

Harding, A. & N. *Victorian Staffordshire Figures 1835-1875: Book Three*. Atglen, Pennsylvania: Schiffer Publishing Ltd., 2000.

Harris, Dee, and Jim, Kaye Whitaker. *Josef Originals: Charming Figurines*. Atglen, Pennsylvania: Schiffer Publishing Ltd., 1999.

Hastins, Bud. *Bud Hastin's Avon & C.P.C. Collector's Encyclopedia: The Official Guide for Avon Bottle Collectors (15th Edition)*. Collectors Pub., 1998.

Heiremans, Marc. *Art Glass from Murano*.

Heritage, Robert J. *Royal Copenhagen Porcelain: Animals and Figurines*. Atglen, Pennsylvania: Schiffer Publishing Ltd., 1997.

Hockenberry, Dee. *Steiff Bears and Other PlaythingsPast and Present*. Atglen, Pennsylvania: Schiffer Publishing Ltd., 2000.

Horowitz, Joseph. *Figural Humidors: Mostly Victorian*. PTJ Publishers, 1998.

Hunting, Jean & Franklin. *The Collector's World of Inkwells*. Atglen, Pennsylvania: Schiffer Publishing Ltd., 2000.

Husfloen, Kyle. *Antique Trader's Antiques & Collectibles Price Guide 2000*. Antique Trader, 2000.

Huxford , Sharon and Bob. *Schroeder's Collectible Toys: Antique to Modern Price Guide 2000, 6th Edition*. Collector Books, 1999.

Huxford , Sharon and Bob. *The Collector's Encyclopedia of Brush-McCoy Pottery: Updated Values, 1996 Values*. Collector Books.

Jacobs, Sharon. *Collector's Guide Stringholders*. L-W Publishing & Book Sales, 1996.

Jaffe, Alan. *J. Chein & Co.: A Collector's Guide to an American Toymaker*. Atglen, Pennsylvania: Schiffer Publishing Ltd., 1997.

Johnson, Donald-Brian, Leslie Piña. *Chase Complete: Deco Specialties of the Chase Brass & Copper Co*. Atglen, Pennsylvania: Schiffer Publishing Ltd., 1999.

Jones North, Jacquelyne Y. *Perfume Cologne & Scent Bottles*. Atglen, Pennsylvania: Schiffer Publishing Ltd., 1999.

Katz-Marks, Mariann. *The Collector's Encyclopedia of Majolica*. Collector Books, 1996.

Kenny, Adele. *Staffordshire Animals*. Atglen, Pennsylvania: Schiffer Publishing Ltd., 1998.

Klug, Ray. *Antique Advertising Encyclopedia*. Atglen, Pennsylvania: Schiffer Publishing Ltd., 1999.

Korbeck, Sharon and Elizabeth Stephan. *Toys and Prices 2000*. Krause Publications, 1999.

Kovel, Ralph M. and Terry H. Kovels' *Antiques & Collectibles Price List 2000*. Crown Publishing, 2000.

Kovel, Ralph M. and Terry H. Kovels' *New Dictionary of Marks: Pottery and Porcelain, 1850-Present*. Crown Publishing, 1986.

Kovel, Ralph M. and Terry H. *Kovels' Quick Tips : 799 Helpful Hints on How to Care for Your Collectibles*. Crown Publishing, 1995.

Kuritzky, Louis. *Collectors Guide to Bookends, Identification and Values: Identification and Values*. Collector Books, 1997.

Langham, Marion. *Belleek Irish Porcelain: An Illustrated Guide to over Two Thousand Pieces*. Radnor, Pennsylvania: Wallace Homestead, 1993.

Lindenberger, Jan. *Collectible Ashtrays*. Atglen, Pennsylvania: Schiffer Publishing Ltd., 2000.

Lindsay, Irene and Ralph. *ABC Plates & Mugs Identification and Value Guide*. Collector Books. 1998.

Luckey, Carl F. *Luckey's Hummel Figurines & Plates: Identification and Value Guide 11th Edition*. Books Americana, 1998.

Mahoney, Jeff. *Collectors Value Guide Ty Beanie Babies: Collector Handbook and Price Guide Winter 1999*. Collectors Publishing, 1998.

Mangus, Bev and Beverly and Jim. *Collectors Guide to Banks Identification & Values: Pottery, Porcelain, Composition*. Collector Books, 1998.

Mashburn, J. L. *The Postcard Price Guide: A Comprehensive Reference*. Colonial House, 1997.

McBride, Gerald P. *A Collector's Guide to Cast Metal Bookends*. Atglen, Pennsylvania: Schiffer Publishing Ltd.,

McCarthy, Ruth. *Lefton China*. Atglen, Pennsylvania: Schiffer Publishing Ltd., 1998.

McPherson, Linda. *Modern Collectible Tins Identification & Values: Identification & Values*. Collector Books, 1998.

Megura, Jim. *Official Price Guide to Bottles, 12th edition*. House of Collectibles, 1998.

Norfolk, Elizabeth, Miller (Ed.), Judith (consultant). *Miller's International Antiques Price Guide*. Great Britain: Miller's, 1999.

Moore, Andy & Susan. *The Penny Bank Book: Collecting Still Banks*. Atglen, Pennsylvania: Schiffer Publishing Ltd., 2000.

Murray, *Wade Whimsical Collectibles, 3rd Edit*. Charlton Standard Catalogue. 1996-97 Values.

Nickel, Mike, and Cindy Horvath. *Kay Finch Ceramics: Her Enchanted World*. Atglen, Pennsylvania: Schiffer Publishing Ltd., 1996.

Paradis, Joe & Joyce. *The House of Haeger 1914-1944: The Revitalization of American Art Pottery*. Atglen, Pennsylvania: Schiffer Publishing Ltd., 1999.

Perkins, Fredda. *Wall Pockets of the Past: Collector's Identification and Value Guide*. Collector Books, 1995

Pina, Leslie. *Fifties Glass*. Atglen, Pennsylvania: Schiffer Publishing Ltd., 2000.
Polak, Michael. *Bottles: Identification and Price Guide 2nd Edition*. Avon Books, 1997.
Riebel, Jim. Sanfords *Guide to Nicodemus*. Adelemore Press, 1998
Righini, Dr. Fernando, & Marco Papazonni. *The International Collectors' Book of Cigarette Packs*. Atglen, Pennsylvania: Schiffer Publishing Ltd., 1998.
Robak, Patricia. *Dog Antiques & Collectibles*. Atglen, Pennsylvania: Schiffer Publishing Ltd., 1999.
Roberts, Brenda. *The Collectors Encyclopedia of Hull Pottery*. Collector Books, 1997 Values.
Roerig, Fred and Joyce. *Collector's Encyclopedia of Cookie Jars, Books 1 and 2*. Collector Books, 1997.
Roller, Gayle. *Hagen-Renaker (2nd edition):* The Charlton Standard Catalogue. The Charlton Press, 1999.
Rontgen, Robert E. *The Book of Meissen*. Atglen, Pennsylvania: Schiffer Publishing Ltd., 1996.
Sampson, Shirley B & Irene J. Harris, *Beautiful Rosemead*. Sanders Printing Co., 1986.
Sasicki, Richard and Josie Fania. *The Collector's Encyclopedia of Van Briggle Art Pottery: An Identification & Value Guide*. Collector Books, 1995 Values.
Schneider, Mike. *California Potteries: The Complete Book*. Atglen, Pennsylvania: Schiffer Publishing Ltd., 1995.
Schneider, Mike. *Royal Copley: Identification and Price Guide*. Atglen, Pennsylvania: Schiffer Publishing Ltd., 1995.
Schneider, Mike. *Stangl and Pennsbury Birds: Identification & Price Gui*de. Atglen, Pennsylvania: Schiffer Publishing Ltd., 1994.
Schneider, Mike. *Animal Figures with Price Guide*. Atglen, Pennsylvania: Schiffer Publishing Ltd., 1990.
Schneider, Mike. *Ceramic Arts Studio: Identification and Price Guide*. Atglen, Pennsylvania: Schiffer Publishing Ltd., 1994.
Schneider, Stuart and George Fischler. *Cigarette Lighters*. Atglen, Pennsylvania: Schiffer Publishing Ltd., 1996.
Schroy, Ellen T. (Ed.). *Warman's Antiques and Collectibles Price Guide 34th Ed.* Warman Publishing Co., 2000.
Scott Publishing. *Scott Standard Postage Stamp Catalogue*, Sidney, Ohio. Scott Publishing Company, 1995.
Seecof, Robert & Donna, Louis Kuritzky. *Bookend Revue*. Atglen, Pennsylvania: Schiffer Publishing Ltd., 1996.
Sieber, Mary. *1999 Price Guide to Limited Edition Collectibles (Serial), 4th edition*. Krause Publications, 1998.
Sillar, F. C. & R. M. Meyler, *Elephants Ancient and Modern*, Viking Press, 1968.
Snyder, Jeffrey B. *Canes: From the Seventeenth to the Twentieth Century*. Atglen, Pennsylvania: Schiffer Publishing Ltd., 1993.
Snyder, Jeffrey B. *Hull Pottery: Decades of Design*. Atglen, Pennsylvania: Schiffer Publishing Ltd., 2001.
Spain, David. *Collecting Noritake A to Z: Art Deco & More*. Atglen, Pennsylvania: Schiffer Publishing Ltd., 1999.
Stephan, Elizabeth and Richard. *O'Brien Collecting Toys: Identification and Value Guide, 9th Edition*. Krause Publications, 1999.
Stoddard, Tom and Loretta. *Ceramic Coin Banks: Identification & Value Guide*. Collector Books, 1997.
Storino, Louis. *Chewing Tobacco Tin Tags 1870-1930*. Atglen, Pennsylvania: Schiffer Publishing Ltd., 1995.
Summers, Bobby. *Value Guide to Advertising Memorabilia, 2nd Edition*. Collector Books, 1998.
Terranova, Jerry, and Douglas Congdon-Martin. *Antique Cigar Cutters & Lighters*. Atglen, Pennsylvania: Schiffer Publishing Ltd., 1996.
Tompkins, Sylvia, and Irene *Thornburg. America's Salt & Pepper Shakers*. Atglen, Pennsylvania: Schiffer Publishing Ltd., 2000.
Warner, Ian and Mike Posgay. *The World of Wade: Collectable Porcelain and Pottery Revised edition*. Antique Publications, 1997.
Waterbrook-Clyde, Keith & Thomas. *The Decorative Art of Limoges: Porcelain & Boxes*. Atglen, Pennsylvania: Schiffer Publishing Ltd., 1999.
White, Carole Bess. *Collectors Guide to Made in Japan Ceramics: Identification & Values Books I, II and III*. Collector Books, 1998.
Wood, Jane. *Collector's Guide to Post Cards*. L-W Publishing & Book Sales, 1995.
Zimmerman, David. *Encyclopedia of Advertising Tins 2nd edition*. Collector Books, 1998.

Miscellaneous Resources

American Topical Association: ATA Central Office, Box 50820, Albuquerque, NM 87181-0820. They have list of most elephant stamps, including elephant watermarks.
Brown, Mitch, Elephant Castle Gazette: Pachyderm Collectors News.(email: Elephldy@aol.com). Nevada.
Couden, Melinda. Luck "E" Penny's Hitchhiker's Guide to Elongated Coins.
Indianapolis Zoo. Email: mcouden@netusa1.net, Luck "E" Penny P. O. Box 1511 Beltsville, Maryland 20704-1511.
Enchanted Elephant Collectors Club, 12650 Overseas Hwy. Arathon, FL. 33050.
Huegel Joan L. Jumbo Jargon, 1002 West 25th Street, Erie, Pa 16502-2427.
McKenney, Dorothy. Elephant Collectors Club, P. O. Box 680565, Franklin, TN 37068.

Web Sites

The following table contains a list of web sites I found useful in researching information about elephant collectibles. Although each of the sites existed at the time I visited it, at the address indicated, there is no guarantee that the site still exists. On many of these sites you need to navigate, or further narrow your search to a specific collectible or "elephant" item. Or, you go to the root site instead of the specific page indicated - delete everything after the first slash ("/"). Also, you can to a major search portal like google.com or yahoo.com and type in "elephant" and spend hours and days and weeks and.... looking at elephant stuff.

Site	URL Address
ABC's of Selling Art: Art Buys Classifieds	www.artbuys.com/price_guide.html
Absolute elephant links	www.elephant.se/links.htm
Absolutely Vintage-Vintage Jewelry	www.absolutelyvintage.net
Allaboutart	www.allaboutart.com
Amazon.com: type in "elephant"	www.amazon.com
Amphora and Teplitz	www.tias.com/stores/vintageinc/
Porcelain_and_Pottery-Amphora information from The Zsolnay Store	www.drawrm.com/amphinfo.htm
Anthony's-Coins, Stamps	astampcoin.com
Antique Toy Company Information	www.antiquetoys.com/companies
Antique Vocabulary	www.leitzgal.com
Antiques and Collectibles	www.fantiques.com www.collectoronline.com
Antiques and Decorative Items Catlg	www.scarlettsweb.co.nz
Antiques at The Drawing Room of Newport, RI & The Zsolnay Store	www.ccantiques.com
Antiques stores	www.directory.teradex.com/
Art Deco & Art Nouveau	www.collectics.com/deco_nouveau.html
Art Deco	www.bookmallventura.com/artdeco.html
Art Reproduction	www.djvart.com/index.asp
Artesania Rinconada	www.rinconada.com
Artist & Designers.	www.empiregiftsandantiques.com/Artists/index.cfm
Artist Galleries	secure.shopaac.com/aashop/galleries
Artist gallery with Boulanger material	www.eslawrence.com
Arts & Crafts Movement Metalwork	www.Suite101.com
Asian art	www.asiaartcollection.com
Baccarat crystal	www.antonio-gib.com/nfbaccarat.htm
BangkokNet Asian Elephant Giftshop	www.bangkoknet.com/elephant/giftshop.html
Bargain Shack offers Boyd Glass, Mosser	www.bargainshack.com/index.html?catalog9_0.html
Barrett-Smythe Animal Zippo Lighters	www.integracom.net/smoker/BS/bspewter.htm
BEST Antiques & Collectibles Sites on the Web!	computrends.com/antiquering.html
Blewster's Antique Flow Blue	www.simons.net/blewsters
Blue Elephant Wares Doulton, Beswick	www.blueelephantwares.com
Bob's Rock Shop: Mineral Images	www.rockhounds.com/rockshop/minimage.html
Boyd Glass Factory Pressed Colored Glassware	www.boydglass.com
Boyd Glass from the Glass Encyclopedia	www.encyclopedia.netnz.com/Boydglass.html
Bronze & Silk Sculptures	www.silk-elephant.com/art/bronze-statues.html
Bronze Outlet - Fine Bronzes	www.bronzeoutlet.com
Bronze Shop - Distinctive Lamps	www.bronzestatue.com/lamps.html
Bullyland Prehistoric Mammals	www.dinofarm.com/bulprehmam.html
Butterfield & Butterfield	www.butterfields.com
California Pottery, Kay Finch, Brad Keeler, etc.	www.webcases.com/cal-pot1.html
Canadian Pottery Identifier	www.angelfire.com/nc2/canadianpotteryid
Capel Collection	www.seriouscollector.com/dealers/CapelCollection.htm
Capell, Ted Elephant collectibles & info.	www.elephanthunter.com
Carl Sorensen Metal Work	www.the-forum.com/silver/soren.htm
Carving galleries Resources	www.world-arts-resources.com/galleries/auto/Carving.html
Ceramics & Pottery Links	www.freepages.ugo.com/murphysclay/Ceramics_Pottery_Links.html
Ceramics research info	www.bcsd.k12.ca.us/stiern/library/links2/ceramic.html
Chess Gallery	www.execulink.com/~starmap/chess/gallery.html
China & Pottery - Antiques Catalogue	www.scarlettsweb.co.nz/catalog
Christine's Closet-Limoges	www.chrissy.com/limoges.html
Collectibles Classifieds, Auctions	www.collectit.net
Collectibles superstore	www.rhodas.com
Collectibles	www.collect.com
Collectibles	www.icollector.com
Collect-Online Websites: Ceramics Collectibles	www.collect-online.com/links2/pages/Ceramics
Collector Online #152: Antiques & collectibles	www.collectoronline.com/booths/booth-152/3589.html
Collector Online Booth 47: Dan & Nancy's Antiques	www.collectoronline.com/booths/booth-47/1556.html
Collector Online's Club Directory	www.collectoronline.com/club-directory.shtml
Collector's Info.	www.collectorsinfo.com/newsletter/newsletter.html
Country Artists Tuskers	www.countryartists.co.uk/site/indext.htm
Country Victorian Mag-Wedgwood	www.countrycollector.com/victorian23/

wedgewood.html
d'Accueil Baccarat www.baccarat.fr/baccarat
Dan Ryder Fossils/Paleoguy Online www.paleoguy.com
Daum www.magnon-jewelers.com/Other/daum.htm
Dedham Pottery www.dedhampottery.com
Delft www.delftsepauw.com
Depression Glass for Sale www.marys-antiques.com
Diecast Toys A to Z www.toynutz.com/a-z.html
Discount Knives & Swords www.knifecenter.com
Doris Stephan's Antiques & Collectibles www.stephan.net/dds/othx.html
Douglas Van Howd www.douglasvanhowd.com
Doulton www.blueelephantwares.com/Doulton.html
Eagle Ridge Collectibles www.tias.com/stores/erc/
East European Art Deco Artists www.allaboutdeco.com/East%20European.htm
EBAY elephants for auction search-desc.ebay.com/search/
Elephant Artifacts Sources www.wildheart.com/sources/main_sources.html
Elephant Carved in Stone www.stonecarver.com/gargoyle.html
Elephant Collectible Index www.someonespecial.com/
Elephant Collectibles www.targetexpress.com/html/elephants.html
Elephant Figurine Collectibles-Tuskers www.artglass-pottery.com/
elephant.htm
Elephant Gallery www.hrgallery.8m.com/index
Elephant Gallery www.wakhok.ac.jp/zou/elephantG.html
Elephant Gifts www.easternorigins.com/elephants.php3
Elephant Head Knife, Franklin Mint www.wildheart.com/sources/30_franklin_knife.html
Elephant info & ads w/elephants www.elephanteria.com/
Elephants for sale www.dlcs.com/elephants/
Environment News:
Africa's First Peace Park ens.lycos.com/ens/apr99/1999L-04-16-03.html
Fenton Glass www.encyclopedia.netnz.com/Fentonglass.html
FredaLA-Web's Collection fredala.com/wedelfromnoa.html
Gamini Ratnavira noted wildlife artist www.gaminiratnavira.com
Gerhard Skrobek www.kathieschristmas.com/gerhard_skrobek.htm
Getanimals.Com www.getanimals.com
Giant Woolly Mammoth www.dinofarm.com/gianwoolmam.html
Gifts www.cannylink.com/shoppinggifts.htm
Glass Animals from B & B SHOP www.facets.net/bbshop/animal.htm
Glass Cases & Domes www.super-highway.net/users/deroberts/
glasscases.html
Glass Museum On Line
from Angela Bowey www.glass.co.nz
Glass notes: manufacturers, A to D www.great-glass.co.uk/mana-d.htm
Gold Coins: Phoenix Enterprises www.nomisma.com/phoenix1.htm
Gothic furnishings, antique textiles, etc. www.drawrm.com
Graves Son and Pilcher www.cm-net.com/ai/graves/auct/a970918
Hen on the Nest Mfgs home.att.net/~jeanettem/manufac.htm
Herend,Porcelain Figurines & China www.giftcollector.com/html/herend
Heubachs www.stanecdolls.com/TechnicalData.htm
Hindu indiaserver.com
Hollohoza USA www.Hollohaza.com
I.M. Chait Monthly Auction Catalog www.chait.com
Import collectibles www.cyberimport.com/catalog/index.htm
Imported Hand Carved Elephants www.handcarvedelephants.com
India Six O www.indiasixo.com/items.html
Inspired Planet www.inspiredplanet.com/catalog.html
Internet Antique & Collectible Mall www.tias.com
Intl. Soc. of Antique Scale Collectors www.collectoronline.com/clubs/ISASC
Ivory source for artists www.ivoryworksltd.com
J. & A. Edwards Ltd www.edwardschina.co.uk/acatalog/
Jack Bryant bronze wildlife sculptures www.jackbryantjrart.com
Jade and Stone Animal Carvings www.chadanco.com
Jewelry www.qconline.com/dusty/Direct.htm
www.theplace2b.com/jewelry/
Jiang posters www.postershop.com/Jiang-p.html
JL Gallery Arts ETC bronzes www.jlgalleryartsetc.com/clark.htm
JP Crystal Collectibles: Lladro,
Swarovski, etc. www.jpcrystal.com
K's Korner On-Line Gifts www.pages.prodigy.net/robinski/index.html
Kovels www.kovels.com
LA Modern Auctions www.collectiblesworld.net/community/newstand/
la.htm
Lemley's Knives www.lemleys.com/ind.html
Lenore Atwood elephant artist www.interlog.com/~lenore
Lenox Collections www.lenoxcollections.com
Lenox sites www.lennox.com
Lenox sites www.lennoxcollections.com
LeRoy Neiman Website www.leroyneiman.com
Links to Antique sites www.frontiertimes.com/links/antiques.html
Lisa's Rock Shop www.minermike.com
Lladro - Horses & Elephants www.someonespecial.com
Lladro Figurines www.allensinc.com/lladros/1150.htm
Lost Wax Glass by Rolanda Scott,
Liz Marks, etc. www.artglass-pottery.com/artglass7.htm
Louisiana Antique Lighting www.antique-center.com/louisiana/
antique_lighting.htm
Metzeler Online www.metzeler.com
Mineral Treasures www.members.tripod.com/~l_young/index.html
MJL Catalog www.cliq.com/clients/mjl/MJL_cat.html
Murano glass animals www.anabays.com/products/animals1.html
Murano Glass www.store.yahoo.com/museumcompany/
5000132.html
Nao by Lladro. Nao Figurines www.store.yahoo.com/bestwishes/naobylladro.html
Nature's Emporium Index www.natures-emporium.com/ind.html
Nautilus Arts and Crafts Homepage www.nautilus-crafts.com
Old Sellwood www.oldsellwoodantiquerow.com/directory.html
Olifant Citron Vodka www.ivodka.com/olifant-citron.html
Omnibus Sandcastle Antiques sandcastleantiques.com
Online collectors clubs www.collectorsonline.com
Opalescent Glass www.glass.co.nz/opalesc.html
Opium Weights, Ancient Bronze mighty-bead.com/index.html
oster gallery www.postergallery.com/list.asp
Paul's Elephant Collectibles www.geocities.com/eureka/concourse/1930/
index.html
Phillips-auctions home www.phillips-auctions.com/index.html
Places To See Elephants: Central USA www.wildheart.com/sights_sites/
sights_usa.html
Poetry in porcelain www.the-hindu.com/2000/10/19/stories/
0419401s.htm
Porcelain Dolls - Heubach Seminar www.stanecdolls.com/heubachdollsseminar.htm
Prince Bernhard background www.bilderberg.org/bernhard.htm
Prince Bernhard www.britannica.com/bcom/eb/article/3/
0,5716,80983+1+78853,00.html
Prince Bernhard www.espu-trust.org/Prince_Bernard.htm
Rare Elephant Pillow Covers www.international-e-market.com/Home-Interior/
Elephant-pi
Reproduction news www.repronews.com/indexpage.html
Royal Crown Derby - Paperweights www.royal-crown-derby.co.uk/pw_list.html
RubyLane antiques mall www.rubylane.com
Sabino Art Glass wwwsabinoartglass.com
Sabino Glass www.encyclopedia.netnz.com/Sabinoglass.html
Sacred Buddhist Painting -
The Tibetan Thangka www.exoticindiaart.com/thangkas.htm
Salzburg Creations: bronzes www.salzburgcreations.com
Selling gemstone jewelry, etc. www.colorsoftheorient.com
Silk Road Trading Concern www.silkroads.com
Silver Spoons & Salt Dips at
Christy's Gifts www.christysgifts.com
Silver/CN Coins: Phoenix Enterprises www.nomisma.com/phoenix2.htm
Silverware - Sterling and Silverplated pw2.netcom.com/~flatware/index.html
Simon Andrews Stamps - Home www.simonandrews.com
Snap-Shot -
Animal Pictures & Wallpaper www.snap-shot.com/frame/animals.htm
Special Collections www.wakhok.ac.jp/zou/elephant/elephant.html
Steals & Deals Page Three www.absolutelyvintage.net/steals3.htm
Stonetrade: Soapstone Fountains, etc. www.stonetrade.com/Decor.html
Sylvac china, figurines, etc. www.208.186.123.39/shopworld/sites/oadby/
online_store/
pages/Sylvac.html
Sylvac www.sylvac.co.uk/HTM/sylvachome.htm
Teplitz & Amphora @ Collectics
Virtual Museum www.collectics.com/museum_teplitz.html
Thai Animals www.thaianimals.com/search.asp
Tokens/Medals: Phoenix Enterprises www.nomisma.com/phoenix4.htm
Trade tokens- NATCA www.flash.net/~tokenguy/index.htm
Treasure Traders, Bahamas.
Large Herend collection www.treasuretraders.com
Treasures from a Bygone Era www.treasurescatalog.com/cgi-bin/treasures.shop
Trocadero-Antiques, Collectibles www.trocadero.com/directory
Troll Company www.surf.to/dam.
USGS Photo Glossary www.volcanoes.usgs.gov/Products/Pglossary/
obsidian.html
Van Briggle Art Pottery, Ceramics www.vanbriggle.com
Victorian Price Guide by
Slawinski Auctions www.slawinski.com/price_guide/g_page11.htm
Vintage Ronson Lighters members.tripod.com/~Transporter/index.html
Walden's Antique and
Collectible Art Glass www.matchings.com
Wedgwood www.wedgwood.com
World of Products www.valuenetwork.com/sab/busdir/genbs/
to.on.ca.wop.html
World Wildlife Endangered
Species Coin Collection www.rimart.com/coin.html
World-wide Artists www.novica.com
Zack the Boyd elephant www.boydglass.com/lists/zack.htm
Zippo Company Profile www.zippo.com/standard/about/profile/index.html
Zippo Lighters www.theconnections.com/collectibles1.html
Zippo lighters, butane lighters, & pipes www.pipeshop.com

Section 5

Index of Manufacturers and Artists

The following index contains manufacturer and artist names, along with a few other words (e.g., famous elephant names like Jumbo, Dumbo). Together with the detailed table of contents, you can easily find the elephants you want to see. Although I could have included words like pottery, jewelry, and advertisement they would be too broad for finding specific elephants fast.